THE BEGINNER'S GUIDE TO
SELF SUFFICIENCY PROJECTS FOR THE HOME

Quarto.com

© 2024 Quarto Publishing Group USA Inc.
Text © 2017 Quarto Publishing Group USA Inc.

First Published in 2024 by New Shoe Press, an imprint of The Quarto Group,
100 Cummings Center, Suite 265-D, Beverly, MA 01915, USA.
T (978) 282-9590 F (978) 283-2742

All rights reserved. No part of this book may be reproduced in any form without written permission of the copyright owners. All images in this book have been reproduced with the knowledge and prior consent of the artists concerned, and no responsibility is accepted by producer, publisher, or printer for any infringement of copyright or otherwise, arising from the contents of this publication. Every effort has been made to ensure that credits accurately comply with the information supplied. We apologize for any inaccuracies that may have occurred and will resolve inaccurate or missing information in a subsequent reprinting of the book.

> **Essential, In-Demand Topics, Four-Color Design, Affordable Price**
> New Shoe Press publishes affordable, beautifully designed books covering evergreen, in-demand subjects. With a goal to inform and inspire readers' everyday hobbies, from cooking and gardening to wellness and health to art and crafts, New Shoe titles offer the ultimate library of purposeful, how-to guidance aimed at meeting the unique needs of each reader. Reimagined and redesigned from Quarto's best-selling backlist, New Shoe books provide practical knowledge and opportunities for all DIY enthusiasts to enrich and enjoy their lives.
>
> Visit Quarto.com/New-Shoe-Press for a complete listing of the New Shoe Press books.

New Shoe Press titles are also available at discount for retail, wholesale, promotional, and bulk purchase. For details, contact the Special Sales Manager by email at specialsales@quarto.com or by mail at The Quarto Group, Attn: Special Sales Manager, 100 Cummings Center, Suite 265-D, Beverly, MA 01915, USA.

10 9 8 7 6 5 4 3 2 1

ISBN: 978-0-7603-9100-6
eISBN: 978-0-7603-9101-3

The content in this book was previously published in *Step-by-Step Projects for Self-Sufficiency* (Cool Springs Press 2017).

Library of Congress Cataloging-in-Publication Data available

Photographer: Tracy Walsh
Photo Assistance: Ian Miller, Susan Storck
Illustrator: Christopher R. Mills Illustration (pages 14, 114)

Printed in China

THE BEGINNER'S GUIDE TO SELF SUFFICIENCY PROJECTS FOR THE HOME

Grow Edibles, Raise Animals, Live Off The Grid & DIY

EDITORS OF
COOL SPRINGS PRESS

Contents

Building A Self-Sufficient Life 6

SECTION 1: CHICKENS AND OTHER CREATURES 8
    01 Chicken Coop 10
    02 Brooder Box 20
    03 Large Farm Animals 24
    04 Top-Bar Beehive 28

SECTION 2: NOURISHING YOUR GARDEN 34
    05 Compost Bin 36
    06 Two-Bin Composter 44
    07 Collecting Rainwater 50
    08 Channeling Rainwater 56

SECTION 3: GARDEN PROJECTS 62
    09 Starting and Transplanting Seedlings 64
    10 Clothesline Trellis 70
    11 Pallet Planter 76
    12 Raised Beds 80
    13 Cold Frame 88
    14 Planting Trees 94

SECTION 4: FOOD PREPARATION AND PRESERVATION 98
    15 Preserving Your Bounty 100
    16 Solar Dryer 108
    17 Root Vegetable Rack 112

SECTION 5: HOMESTEAD AMENITIES 118
    18 Solar Oven 120
    19 Frame Loom 128
    20 Manual Laundry Washer 134

Resources 140
Credits 140
Index 141

Building a Self-Sufficient Life

Self-sufficient living is a highly complementary practice—once you begin, you'll find that many parts of your home are connected, and that multiple systems of self-sufficiency contribute to one another, often corresponding with the natural cycles of the earth.

Because of this interconnectedness, many of the projects in this book will naturally lead you to more and more projects that will help you maximize your self-sufficiency work. For example, if you start a garden, the fruits and vegetables you grow will provide waste that will transform into the compost that will nurture next year's bounty. Setting up a rainwater collection system not only reduces your reliance on public utilities, but the fresh, soft water will also help your plants grow healthy. The hens you are raising for their eggs control garden pests and provide free fertilizer.

At the end of the growing season, you'll likely be overwhelmed with vegetables and will need to find a means to store and preserve them—perhaps a basement root cellar could be a good option. And, by growing organic vegetables nurtured by compost and animal manure, you create a pesticide-free habitat for honeybees to prosper, while they, in turn, pollinate the plant life.

That said, you do not need to take on all the projects in this book at once. Start with the projects that naturally supplement the efforts your family is already making toward self-sufficiency. If you already recycle, a natural next step is to build compost bins and begin to make compost with food and paper waste as well. If you already maintain a beautiful flower garden, why not build a home for the honeybees that are already frequent visitors, allowing you to collect the honey they produce? If you already grow fruits and vegetables, why not build a solar fruit dryer or a drying rack, or, if you have apples, make a cider press? If you already garden, why not build a greenhouse?

For the newcomer, the projects on the following pages provide multiple opportunities to create a more self-reliant lifestyle. For the experienced self-sufficient homeowner, the step-by-step projects included here will provide you with the means to expand and streamline your efforts.

SECTION 1

Chickens and Other Creatures

A big part of self-sufficiency is producing your own food. But as delicious and nutritious as fruits and vegetables may be, protein is an important part of every diet. That's where food animals come in, especially chickens. Chickens provide an ongoing harvest of eggs and can supply valuable meat as well.

Beyond food, some living additions to your self-sufficient homestead provide support in different ways. Goats give you milk, which can be processed into cheese. Bees provide a valuable service in pollinating garden plants and supply delicious honey in the bargain—making them some of the most beneficial living things you can raise on your property. They also require very little care and upkeep, making them even more desirable for any homestead looking to become self-sufficient with a minimum of free time.

The projects in this section are all about taking the best care possible of the animals that call your home, well, home. These structures are specifically designed and suited to a given creature, but they have also been designed to look good. These are simple structures—you won't need homebuilding experience to put up the chicken coops we outline in the pages that follow, nor will you need to be a master woodworker to assemble your own honey-making beehive.

With a little bit of work and a modicum of tools and materials, you can make homes within your homestead for living things that will help you live more self-sufficiently.

01 | Chicken Coop

This chicken coop is just over 4 feet by 8 feet, with a 6-foot-wide roof, so it won't take up much space in your yard. It's big enough to comfortably house up to six chickens, and they'll stay dry and safe from predators under the large roof. Two people (or one strong person with a dolly) can easily roll the coop around the yard, and in the winter the plywood enclosure is tight enough to keep the chickens comfortable.

The framing for the coop has been keep to a minimum so that it's easier to move and quicker to build. We've also used 2 × 2s instead of 2 × 4s as much as possible to lighten the weight. Once you have all the materials on hand, you can build the coop in a day. If you have more chickens to house, the basic plan can easily be expanded—just keep the width and height the same and extend the length so that you can make economical use of 4 × 8 sheets of plywood.

The coop is large enough for two or three nesting boxes, which you access by opening the large door in front and leaning in. These can be as simple as plastic crates. The chickens will also appreciate a roost inside the coop. Make this from a closet rod pole or a length of 2 × 2, placing it about 18 inches up and running it from wall to wall in the middle of the coop. A window opening under the eave in the back provides ventilation, even in heavy rain. In the winter, simply cover it with plexiglas or plastic. You can also look into simply installing a small, inexpensive, utility-grade window in one of the sides—a bit luxurious for chickens, but it will give you an easy way to regulate the temperature and airflow inside the coop.

House your chickens in style in this large, movable coop. The chickens will appreciate the extra space, and the wide roof helps keep the ground from getting muddy.

Chickens are outdoor birds and prefer to roam within limits. They are natural pest-controllers and will stay healthiest if allowed to move about for a portion of every day.

Chickens spend most of their day foraging for bugs and tasty bits of greenery, then return to their coop every night, even if they've been wandering in the yard.

Getting Started

If you plan to raise chickens for their eggs, you can buy them either as newborn chicks or as pullets (about four to six months old). Chickens usually start laying at about six months. However, the older the chickens are, the more they'll cost. Another point to keep in mind is that handling baby chicks when they're young makes it easier to handle them as adults—but also harder to butcher for meat. Do an online search for chicks for sale, or talk to local dealers for more information. Pullets do not require a special brood environment, as chicks do (see Brooder Box, page 20), but you should monitor their light exposure and heat when they're young. Keep your pullets in the coop for a week or so to help them get accustomed to their new home.

When they are old enough, allow your hens out of the coop during the day to peck and wander around a larger enclosed area, such as a small yard surrounded by a fence. Hens will not wander far. They love to dine on the bugs and weeds in your yard, and will produce a greater yield of healthier eggs if allowed to move around freely. At night, make sure all your hens are safely locked in to the coop to sleep.

Collecting Eggs

Collecting eggs from a brooding hen requires a careful hand and sound timing. Expect a good pecking if you reach into the nest while mother is awake. The best time to gently remove eggs from the nest is in the morning or during the night, when hens roost. This is also the best time to pick up a hen and move her, because she won't argue while she's sleeping. Eggs may be brown, white, or sometimes even light blue or speckled—depending on the breed of your chicken. No matter the appearance, what's inside will taste the same.

Gather eggs twice a day, and even more frequently during temperature extremes when eggs are vulnerable. The longer they sit in the nest, the more likely eggs are to suffer shell damage. After gathering, pat the eggs clean with a dry cloth. If they are noticeably dirty, wash them with warm water. Place clean, dry eggs in a carton and refrigerate.

APPROVALS

The first step to starting your own chicken coop is to get permission from your local municipality. Many cities and towns allow homeowners to keep hens, but no roosters. Typically, there is also a limit on the number of hens you can keep, and the distance your coop must be located from your neighbors' windows. Check the regulations in your municipality as you develop your chicken ranching plan. It's also important to talk with your neighbors to seek out their consent, even if their written permission is not required.

CHICKEN BREEDS

So, why do you want chickens?

Perhaps you dream of eggs with thick, sturdy sunshine yolks that are unbeatable for baking (and perhaps for selling at a farmers' market stand). Maybe you want to dress the dinner table with a fresh bird. Not sure? If you want both eggs and meat, you're safe with a dual-purpose breed such as the barred Plymouth Rock.

Next, consider the size flock you will need to fulfill your goals. This depends on land availability and how much produce you wish to gain. In other words, if volume of eggs or meat matters, then you increase your "production line." If your reason for raising chickens is to enjoy the company of a low-maintenance feathered pet—the meat and eggs are just a bonus—then a flock of three or four hens and possibly a rooster will get you started.

Layers

While all chickens produce eggs, laying breeds are more efficient at the job than other breeds; in short, layers lay more eggs. You can expect about 250 eggs per year or more if your layer is more ambitious than most. Laying hens tend to be high-strung, however, and while they lay many eggs, they show little interest in raising chicks. You may reconsider laying breeds if you want your hens to raise the next generation. Layers simply aren't interested—but they'll keep seconds coming to the breakfast table.

Meat Breeds

These chickens are classified based on size when butchered. Game hens weigh 1 to 3 pounds (0.5 to 1.4 kg), broilers (also called fryers) range from 4 to 5 pounds (1.8 to 2.3 kg), and roasters are usually 7 pounds (3.2 kg) or slightly more. You'll find cross-breeds ideal for the backyard, including broiler-roaster hybrids like the Cornish hen or the New Hampshire.

Dual-Purpose Breeds

Larger than layers but more productive (in the egg department) than meat breeds, dual-purpose breeds are the happy medium. Hens will sit on eggs until they hatch, so you can raise the next generation. There are many chickens that fall into this variety, and their temperaments vary. Many dual-purpose breeds are also heritage breeds, meaning they are no longer bred in mass for industry. They like to forage for worms and bugs, are known for disease resistance, and, essentially, are the endangered species of the chicken world.

Chickens raised for meat usually are purchased from a hatchery or a feed store when just a day or two old. Raising them to broiler weight (4 to 5 pounds) takes 6 to 8 weeks. During this time they will consume around 15 pounds (7 kg) of feed.

Ornamental chickens often make good pets. They enjoy human companionship. And they are a fun and visual addition to the yard!

Chickens do well in cold weather as long as they have a sheltered, insulated roosting area and their water supply is not allowed to freeze.

BUILDING A CHICKEN COOP

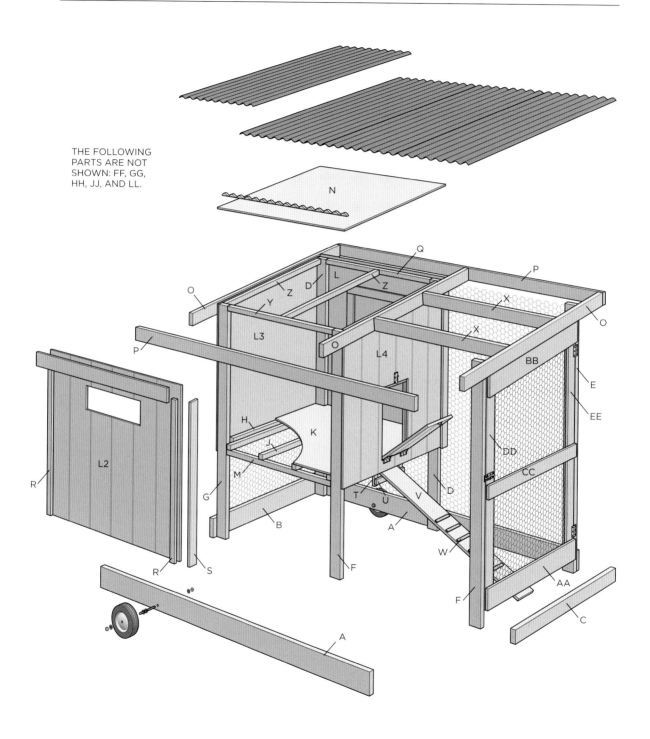

THE FOLLOWING PARTS ARE NOT SHOWN: FF, GG, HH, JJ, AND LL.

CUTTING LIST

KEY	NO.	PART	DIMENSION	MATERIAL
A	2	Base	1½ × 5½ × 93"	2 × 6 P.T.
B	1	Base end	1½ × 5½ × 50"	2 × 6 P.T.
C	1	Base door end	1½ × 3½ × 50"	2 × 4 P.T.
D	2	Front posts	1½ × 3½ × 72"	2 × 4 P.T.
E	1	Front door side post	1½ × 3½ × 72½"	2 × 4 P.T.
F	2	Rear posts	1½ × 3½ × 66"	2 × 4 P.T.
G	1	Rear door side post	1½ × 3½ × 66½"	2 × 4 P.T.
H	2	Floor joist support	1½ × 1½ × 39½"	2 × 2
J	4	Floor joist	1½ × 1½ × 46¾"	2 × 2
K	1	Floor	46¾ × 46¾"	½" plywood
L	1	Front wall	48 × 48 × 58"	T1-11 siding
L2	1	Rear wall	48 × 41¾ × 58"	T1-11 siding
L3	1	End wall	46¾ × 48 to 41¾ × 58"	T1-11 siding
L4	1	Door side end wall	46¾ × 48 to 41¾ × 58"	T1-11 siding
M	2	Floor rim joist	1½ × 1½ × 36½"	2 × 2
N	1	Roof	48 × 48	½" plywood
O	3	Roof beams	1½ × 3½ × 69"	2 × 4 P.T.
P	2	Fascia	1½ × 3½ × 96"	2 × 4 P.T.
Q	2	Top trim	¾ × 3½ × 48"	1 × 4
R	4	Corner trim	¾ × 1½ × 45" (varies)	1 × 2
S	4	Corner trim	¾ × 2½ × 45" (varies)	1 × 3
T	2	Ramp hanger	¾ × 1½ × 6"	1 × 2
U	1	Ramp support	¾ × 1½ × 13"	1 × 2
V	1	Ramp	¾ × 7¼ × 60"	1 × 8
W	14	Ramp battens	¼ × ¾ × 6"	Screen mold
X	2	Roof rafters	1½ × 3½ × 44"	2 × 4 P.T.
Y	2	Roof support	1½ × 1½ × 39½"	2 × 2
Z	3	Roof supports	1½ × 1½ × 44"	2 × 2
AA	1	Bottom rail	¾ × 5½ × 43¼"	1 × 6 cedar
BB	1	Top rail	¾ × 5½ × 43⅝" (angle to angle)	1 × 6 cedar
CC	1	Center rail	¾ × 3½ × 43¼"	1 × 4 cedar
DD	1	Handle stile	¾ × 3½ × 58 1/16" (high edge)	1 × 4 cedar
EE	1	Hinge stile	¾ × 3½ × 64¼" (high edge)	1 × 4 cedar
FF	1	Door stop	¾ × 1½ × 42"	1 × 2
GG	2	Entry door stop	¾ × 2½ × 35½"	1 × 3
HH	1	Entry door stop header	¾ × 2½ × 24"	1 × 3
JJ	2	Ramp door stop	¾ × 2½ × 19½"	1 × 3
LL	1	Ramp door stop header	¾ × 2½ × 16¼"	1 × 2

TOOLS & MATERIALS

- Miter saw
- Circular saw
- Jigsaw
- Drill
- Hammer
- Stapler
- Wrench
- Framing square
- Countersink bit
- ½" wood bit

- Deck screws—1¼", 15⁄8", 3"
- 2½" self-tapping screws or deck screws
- (12) 5⁄16 × 3½" galvanized carriage bolts
- (2) 5½" × ½" galvanized hex bolts with 4 nuts and 8 washers (for tires)
- (4) 1½" galvanized metal angles
- 9 × 1" roofing screws

- (8) 3⁄8" × 4" exterior, self-tapping lag screws
- 4d galvanized casing nails
- (3 pairs) 3" × 3" exterior grade hinges
- (5) galvanized sliding bolts
- 8"-diameter flat-free tires
- 3—2 × 6 × 8' treated
- 10—2 × 4 × 8' treated
- 2—2 × 4 × 10 treated

- 8—2 × 2 × 8' (actual 1½"—untreated)
- 1—1 × 6 × 8' treated or cedar
- 2—T1-11 plywood siding (4 × 8 sheet)
- 1—½" BC plywood
- 6—Green PVC roof panels
- Chicken wire
- Door stop
- Eye and ear protection
- Work gloves

HOW TO BUILD A CHICKEN COOP

Cut the six supporting posts (A, B) to length. Cut the tops at an 8 angle on your miter saw (A, B measurements are to the long edge of the angle). This will translate into a slope of 1½" for every foot.

Cut the 2 × 6 bases and then lay out the sides. Square the posts to the base and then fasten them with 2½" screws (either deck screws or self-tapping screws). The posts will be bolted to the 2 × 6s later, after the plywood sides are fastened—which will ensure that the posts are properly squared, since pressure-treated wood is often not quite straight.

Cut the front and back pieces of plywood siding. Attach the vertical edges of the plywood front (and back) to the two front and back 2 × 4 posts. The plywood should extend ½" past the 2 × 4 on both edges. Fasten the plywood with 1 5/8" screws.

Screw 2 × 2s to the bottom edges of both pieces of plywood siding between the posts. Fasten it with 1 5/8" screws through the face of the siding. Stand the two walls up and join them with 2 × 2s and 2½" screws. Predrill and countersink the screws, driving them at a slight angle starting at about 1¼" from the ends. This will help avoid splitting the ends. Tack on a temporary brace to help keep the walls plumb until the plywood sides are on.

Add two more 2 × 2 joists between the two walls. Cut the plywood floor, notching the corners around the 2 × 4 posts. The plywood should go right to the edge of the posts. Screw the plywood to all the 2 × 2 joists. Also screw additional 2 × 2 joists to the edges of the plywood floor to serve as nailers for the two plywood side panels.

Measure and cut the plywood siding for the angled side at the center of the coop. Attach clamps or temporary blocks to the center posts to hold the piece up as you fasten it. Align one edge of the plywood with the post and fasten it with 1⅝" screws, then align the other edge and fasten it. You may need to do a little pushing, or even shim the base if you're not working on level ground, but aligning this edge will make the coop square and rigid. Leave the other side of the coop open for now to make it easier to complete the framing.

Cut and fasten 2 × 2s to the top edges of all the plywood walls. The angle cuts for the sloping wall are 8.

Bolt the posts to the 2 × 6 bases with 2 carriage bolts each, and then screw on the end pieces (B, C) with ⅜" × 4" exterior, self-tapping lag screws. Unless you're planning to leave the coop in one place permanently, add tires now. We used 8"-diameter flatless tires with ½" bolts for axles.

(continued)

Chickens and Other Creatures

HOW TO BUILD A CHICKEN COOP (continued)

Add a 2 × 2 at the top of the open side of the coop, then screw on a 2 × 2 rafter in the center to help support the roof. Predrill or use self-tapping screws to avoid splitting the wood. Mark the location of this center rafter on the side so you can find it later, then screw on the last piece of siding.

Measure and cut ½" plywood for the roof. Fasten it to the 2 × 2s, aligning it with the edges. The plywood roof is mostly to provide additional warmth in the winter, but if you live in a warm climate, you don't have to install it since it will be covered by roofing panels.

Cut holes in the plywood siding for doors and a window. Use a jigsaw, tipping it into the plywood to get the cut started. Make the cuts as clean as possible—the cut out pieces become the doors. Cut a 20" × 36" main door, a 12" × 20" door for the chickens and a 6" × 20" window. Screw pieces of 1 × 3 around the two doors to serve as door stop, extending them about ½" into the openings.

Cut and install the three roof rafters that run from front to back, making 8° plumb cuts at the front and back. Fasten them to the 2 × 2s inside the plywood with 3" screws. Add two intermediate rafters to support the roof over the open part of the coop.

Finish the roof framing by screwing long 2 × 4 fascias to both sides of the structure. If you're working by yourself, screw a temporary support to one of the end rafters to hold the fascia up. After finishing the roof framing, nail 1 × 4 trim at the top of the plywood between the rafters on the front and back of the coop.

Set the doors into the openings in the siding and attach the hinges and the sliding bolts. Use shims under the doors to elevate them as you put the hinges on. **NOTE:** Some of these hinge and bolt screws may just poke through the inside of the siding. If they do, flatten them out with a file.

The ramp for the chickens combines the back of the small entry door and a longer piece of 1 × 8 that extends from the underside of the coop to the ground. The top end of the 1 × 8 ramp rests on a piece of 1 × wood fastened 2¼" down from the framing (total open area is 2¼" × 8"). When cleaning the open area or moving the coop, just slide the 1 × 8 up and under the framing. Nail small strips of wood for the chickens to hold on to every 4 or 5 inches on the 1 × 8 and the back of the door.

Measure, cut, and assemble outside corners for the coop. Use 1 × 3s for the angled side (8° cut) and 1 × 2s for the back and front sides. Nail on the corners with galvanized casing nails. Staple chicken wire over the inside of the window. In the winter, cover the window with plexiglas or clear plastic.

Assemble a door to the enclosure from 1 × 6s (top and bottom) and 1 × 4s (sides and center). Cut the door to follow the angle of the roof with 8° cuts, and glue and screw all joints. Note that the 1 × 6s are on the outside face of the door. After the glue has dried, staple chicken wire to the back. Fasten hinges near the top and bottom, then attach it to the post, leaving a ½" space at the bottom (note the ½" plywood in the photo) and ⅛" to ¼" of space on the hinge side. Nail a door stop to the handle side.

Attach chicken wire to all the sides with galvanized fence staples or self-tapping, wafer-head screws (wide, flat heads). Fasten the roof panels to the rafters using roofing screws with neoprene washers. Most roof panels come with matching foam or plastic strips that fit the corrugations—insert these along the edge of the roof to seal the gaps and help keep heat in and bugs out. Caulk any inside corners in the coop where you can see light coming through with latex caulk. Throw some hay into the coop and you're ready for the chickens.

02 | Brooder Box

Expectant parents of both human and chicken babies get the same advice: have the nursery ready before you bring home the little one(s). You'll enjoy those precious first weeks a lot more if you're not running around like a . . . well, never mind. For chickens, a nursery is a brooder box, and it's used to provide the same essentials human newborns need: security, warmth, and nourishment. The nice thing about a chicken nursery is that you don't have worry about wall color or getting an ultrasound if you absolutely refuse to go gender-neutral (you're probably expecting all girls anyway).

A brooder box can be literally that simple—just a cardboard box, or a plastic bin, an old fish tank, even a kiddie pool. Backyard farmers often use whatever they have on hand. But a nice, sturdy wood box with a few convenience and safety features will make raising your chicks a better experience, for this brood and many more down the road. This brooder box is made with a single sheet of ¾ inches plywood and measures 36 × 36 × 17⅜ inches, enough room for housing 8 to 10 chicks up to six weeks old. Both the lid and the floor are covered with hardware cloth (wire mesh), making the box secure, well ventilated, and easy to clean.

Heat is provided by a simple clamp-on reflector light, which can be set directly atop the lid's wire mesh or clamped to the light pole at various heights for temperature control. The box lid is hinged in the back and locks in the front with a locking hasp latch. Special removable hinges make it easy to slide off the lid to get it out of the way for a thorough cleaning of the box.

A brooder box like this is a nursery to your newborn chicks, so it's essential that it be comfortable, warm, safe, and nurturing to your future chickens.

BUILDING A BROODER BOX

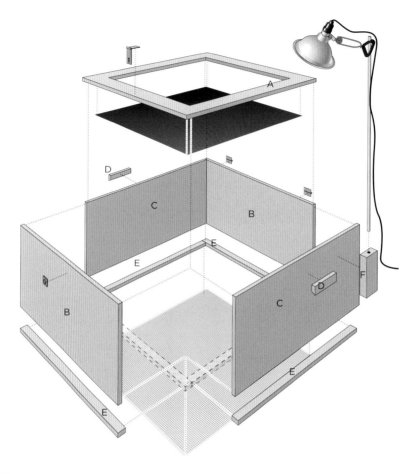

CUTTING LIST

KEY	NO.	PART	DIMENSION	MATERIAL
A	1	Top	¾ × 36 × 36"	¾" plywood
B	2	End	¾ × 36 × 15⅞"	¾" plywood
C	2	Side	¾ × 34½ × 15⅞"	¾" plywood
D	2	Handle	1½ × 3½ × 6"	2 × 4
E	4	Base	¾ × 1½"	1 × 2
F	1	Light mount	1½ × 3½ × 6"	2 × 4

TOOLS & MATERIALS

Circular saw
Drill
Jigsaw
Wire cutters
Staple gun

Wood glue
Sandpaper
Deck screws 1⅝", 2"
½" × ½" galvanized hardware cloth (36 × 70" min.)

Heavy-duty ½" staples
½"-diameter × 36"-long hardwood dowel
(2) Separable lid hinges with screws

Locking hasp latch with screws
Clamp-on light fixture
Zip ties (optional)
Eye and ear protection
Work gloves

The Beginners Guide to Self Sufficency Projects for the Home

HOW TO BUILD A BROODER BOX

Make the lid cutout by marking four lines 4" from the outside edges of the lid piece. Carefully lower your circular saw onto one end of the line and cut to the other end. Finish the cuts at the corners with a jigsaw. **TIP:** Roughly lay out the parts on the plywood sheet before marking and cutting the pieces one at a time.

Assemble the box frame using glue and 2" deck screws. Fit the front and back pieces over the ends of the sides, drill pilot holes, then apply glue and fasten each joint with five evenly spaced screws. Use a framing square to make sure the corners are square as you work. Sand all exposed edges of the box and lid.

Use wire cutters to cut the box bottom and lid cover from ½" hardware cloth. Cut the bottom to 35¾ × 35¾"; cut the lid cover 34 × 34". Screw 1 × 2 cleats to the bottom edges of the plywood box frame, then staple on the bottom cloth with heavy-duty ½" staples. Keep the edges of the mesh ⅛" from the outside edges of the box on all sides. Staple the lid cover to the underside of the lid frame, keeping the mesh 1" from all outside edges. Use plenty of staples on the cover—and staple close to the edges of the cutout—to prevent predators from pushing down the mesh.

Install the box lid using two separable hinges (also called removable or lift-away hinges). Depending on the hinge design, you may want to recess the hinge plates on the box by cutting shallow mortises, using a wood chisel. Add a locking hasp latch to the front side so the lid can lock securely to the box front. Clamp the light fixture to the dowel and plug it in to test the setup.

Create the two handles, beveling one long edge at 15. Cut the strip in half to end up with two 6"-long handles. Install the handles with glue and 1⅝" deck screws so the top edges of the handles are about 2" below the top of the box sides. The long beveled edges face down and toward the box side, creating an easy-to-grab lip for each handle. Inset: Construct the light pole mount by drilling a ½" centered hole in the end of a scrap 2 × 4. Ream the hole until the dowel fits easily and can be slipped in and out to facilitate box cleanings.

03 | Large Farm Animals

Raising farm animals can be a challenge and it is a serious responsibility, but once you've taken the plunge, the rewards can be so great that it is very difficult to go back. But it's important to select animals that will be comfortable on your property.

How hot or cold is the region where you live? While you can make accommodations by choosing the right shelter for your animal, some extreme climates simply are not suitable for every animal. Before you purchase livestock, contact local veterinarians and find out whether they will treat the species you plan to buy. If you cannot find a vet in your area who will treat the llama you're planning to purchase, find out how far you'll have to drive to reach someone who will, and weigh this into your decision. Also, think through the logistics of how you will transport your animal and the time it will take to reach the vet.

Lastly, check with your municipality to find out what the limitations are when it comes to owning livestock. If you live in the city, you may need to request permission from your neighbors to house animals, and some species may not be allowed at all. There may also be setback requirements that apply to animal housing—take these into consideration when planning out the site for your animals' home.

A barn is not necessary for every species, though most animals require some type of shelter for protection from the elements and, perhaps, for sleeping. The type of roof you put over animals' heads depends largely on where you live. In the chilly north, animals will require a more sturdy, draft-resistant abode than in the hot and dry southwest. Build your shelter to suit your climate—in cold climates, your shelter should be designed to keep animals warm, and in warm climates, designed to keep animals cool. Tailor your shelter to meet your animals' needs. For example, pigs require separate space for eating and sleeping, whereas sheep and goats just need a dry shelter to protect them from the elements.

Owning big animals can be rewarding, but they can require lots of work, and you'll need strong fencing to keep them from wandering away and exploring the neighborhood.

Pigs love attention and are infinitely curious. Be sure to build a strong fence around your pigpen to keep your pigs contained.

Pigs

Pigs are personable and intelligent. They also offer a well-rounded learning experience for new animal owners: lessons on the importance of feed mix, daily pigpen cleaning to prevent disease (and smell!), and good old-fashioned recycling. Their composted byproduct is rich fertilizer for your garden.

Pigs don't require a great deal of land, but they do need a dry, draft-free shelter to protect them from the elements. Be sure to prepare this space before bringing home your new pets. You can dedicate a portion of an existing building, such as a barn or shelter, or construct a simple outbuilding. Ideally, the floor should be concrete and sloped for optimal drainage during daily hose-outs. A dirt floor is also fine as long as you replace hay bedding daily.

Pigs also require a separate sleeping and feeding area. A 5 foot by 5 foot square sleeping area will accommodate two pigs. The feed area should be twice this size and contain the feed trough, a watering system, and a hose connection. This serves the double purpose of instant water refills for thirsty pigs and accessibility to your number one pigpen-cleaning tool.

Contain your pigs with a secure fence of woven wire or permanent board. The fence should be about 3 feet tall. Pigs are immensely social animals and love the spotlight, so be sure to spend time with them to help them grow healthy and happy.

Goats

Goats are mischievous class clowns with boundless energy and a lovable nature. They are a great source of both milk and meat, and do not require a vast amount of resources, food, or shelter. They are ruminants that enjoy munching twigs and leafy brush—but if you will be raising goats for their milk, feed them with a forage of hay and grain to preserve their milk's taste. Some goat breeds are also a great source of fibers for fabrics, such as mohair and angora. You'll want to watch these breeds' diets carefully, as their coat will be affected by their diet. Feed them only quality pasture or hay, along with plenty of fresh, clean water.

There's no need to build a fancy house for goats, as they fare well in pretty much any dry, draft-free quarters. Goats are prone to respiratory problems triggered by a moist environment, so avoid heating that can result in condensation. House goats in a three-sided barn, shed, or a shared barn with other animals. It's a good idea to invest your savings on shelter in quality fencing, however. Woven-wire pasture fencing is ideal, and additional strands of barbed or electrical wire will discourage curious goats from escaping.

Goats are high-spirited, lovable, and mischievous animals that love to play. Watch out, though—your goats will try to outsmart you (or your fence!). Goats are a great source of milk, wool, and meat.

Sheep

Sheep are affectionate animals raised most often for their high-quality wool that can be spun into yarn and made into warm textiles. Choose a breed of sheep with a wool density that correlates with the weather where you live—in the north, choose sheep with extra "lining" in their coats; in temperate or arid climates, sheep with fine wool and hair prosper.

When purchasing sheep, look for healthy feet. From the front view, legs and hooves should align, as opposed to being knock-kneed, splayfooted, or pigeon-toed. Check the animal's bite, and be sure there are no udder lumps or skin lesions. Sheep do need to be shorn every spring before the weather heats up, so take good care of your sheep's wool and sell or spin it after it is collected. To care for your sheep (and their wool), make sure they have good nutrition, well-managed pastures, and vaccinations. Sheep need a sturdy fence and some type of shelter, though existing buildings on your land will suit them just fine. They are easy targets for predators, such as coyotes, so make certain your fence is secure.

Lock eyes with a quizzical alpaca and you'll feel like you are being probed for information. Alpacas are gentle animals that are easy to care for and produce soft, extremely warm coats that can be shorn and sold to textile makers.

Alpacas

Alpacas are smaller cousins to llamas and camels and are an approachable, friendly species, which makes them appealing to landowners who want to begin caring for animals. They won't challenge your fencing or trample your pasture. They also require little feed—about a third less than a sheep. Alpacas grow thick coats that are five times warmer than wool and far more durable. Yarn spinners covet alpaca fiber, homeowners admire these loving pets, and investors appreciate the potential returns these valuable creatures promise.

Fencing you'll build for alpacas is designed more to keep predators out than to keep alpacas in. These animals are not ambitious escape artists—not nearly as tricky as goats. But predators can represent a threat to sensitive alpacas, so it's a good idea to install strong perimeter fencing that is at least 5 feet tall. Separate females and males with fencing. Females and their newborns must have separate quarters from the rest of the pack, but do not completely isolate them from the group. A three-sided shelter is adequate for alpacas, which are accustomed to rugged, cold climates. Heat is more of a concern for these animals, and their insulating fiber coats are no help in keeping them cool in summer. A misting system or fans in the alpaca shelter will prevent them from overheating.

Sheep travel in close-knit packs, are spooked easily, and are an easy target for predators, so make sure your sheep are protected by a secure fence.

04 | Top-Bar Beehive

Backyard beekeeping makes more sense today than ever before. Not only are honey bees necessary for pollinating plants and ensuring a better fruit set and bigger crops, they produce delicious honey and valuable beeswax. And recently, the world bee population has experienced a mysterious and concerning dropoff in numbers. Getting homeowners to cultivate a bee colony is a helpful component of the preservation strategy.

In many ways, tending bees is like growing food. There is an initial flurry of activity in spring, followed by ongoing maintenance in the summer, and then harvest in the fall. There is prep work you'll need to complete before you begin and there is a learning curve—you'll need to spend more time with your bees in the beginning until you learn how it's done. Beekeeping is not necessarily an expensive hobby, but with higher-end operations, purchasing the hive, some blue ribbon bees, and all the necessary equipment can require a significant financial investment.

Keeping bees will help you have a better garden, more fruits and vegetables, and honey in the kitchen—even beeswax candles, skin creams, and other natural cosmetics. And, by building a top-bar beehive, you're creating a safe home and enabling one of our earth's most necessary and miraculous species to thrive.

Honey and beeswax are the two commodities a functioning backyard beehive will yield. If your primary interest is honey, build a traditional stacking-box style beehive in which the bees expend most of their energy filling the premade combs with sweet honey. If it's beeswax you seek, make a top-bar hive like the one shown on the following pages.

5 WAYS TO KEEP YOUR BEES SAFE & HEALTHY

1. **Avoid using insecticides in your garden**—Many are long-lasting and toxic to bees.
2. **Buy seeds that are not treated with insecticides**—Some coated seeds may cause the entire plant to become toxic to bees. Check seed packets carefully.
3. **Mix your own potting soil and compost**—Some composts and potting mixes sold at garden centers contain insecticide that is highly toxic to bees and other insects, and will eventually pollute all of your soil. Make your own compost and mix with natural additives for potting plants.
4. **Plant bee-friendly flowers**—Buy wildflower seed mix and plant in uncultivated areas to create small sections of wild, natural habitat for your bees.
5. **Provide a home for bees**—Whether you're a blossoming beekeeper or not, it's easy to provide a home for bees! Provide a simple box as a place for feral bees to nest, or start your own hive.

Top-Bar Hive

Expert beekeeper Phil Chandler insists that beekeeping should be a very simple pursuit, largely because the bees do almost all of the work for you. Chandler, who maintains a website called The Barefoot Beekeeper (see Resources, page 140), is an advocate for natural beekeeping and has designed a top-bar hive that you can build yourself using simple materials. This hive (see pages 31–33) is designed to enable the bees to build their own comb, instead of relying on a premade comb.

The top-bar hive is simple in its construction and, unlike the traditional stacked box Langstroth hive, does not require that you lift heavy boxes to check on your hive's progress, which disturbs the bees within. Rather, you can simply remove the hive roof and inspect the bars one by one without disturbing the rest of the hive. Storage is minimal for a top-bar hive, as there are no supers needed. And, it is not necessary with this hive design to isolate the queen.

This simple top-bar beehive design is a warm and safe home for bees that is easily adjustable to accommodate a growing hive. This design also greatly simplifies the inspection process and minimizes the amount of equipment needed to keep and maintain bees.

BUILDING A TOP-BAR BEEHIVE

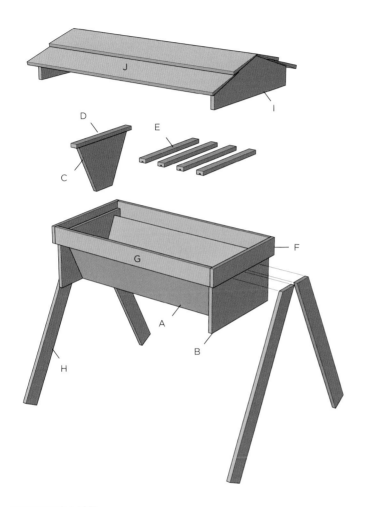

TOOLS & MATERIALS

Lumber (1 × 2, 1 × 3, 2 × 4, 1 × 12)
Carpenter's square
Pencil
Circular saw or table saw
Socket wrench
Exterior-grade construction adhesive
Caulk gun
Clamps
Drill
Tape measure
Hammer
Handsaw
1¼", 2", 2½" deck screws
Stainless-steel or plastic mesh
Roofing nails or narrow crown staples
Eye protection
1" holesaw
⅜ × 2" galv. lag bolts with washers and nuts
Eye and ear protection
Work gloves

CUTTING LIST

KEY	PART	NO.	DIMENSION	MATERIAL
A	Side panel	2	¾ × 11¼ × 36"	Cedar
B	End panel	2	¾ × 11¼ × 19"	Cedar
C	Insert	2	¾ × 11¼ × 15"	Cedar
D	Insert cap	2	¾ × 1½ × 17"	Cedar
E	Top Bar	20	¾ × 1½ × 17"	Cedar
F	Frame end	2	¾ × 3½ × 21"	Cedar
G	Frame side	2	¾ × 3½ × 36"	Cedar
H	Leg	4	¾ × 1½ × 36"	Cedar
I	Cap end	2	¾ × 7¼ × 23"	Cedar
J	Roofing	4	⅝ × 5⅞ × 40"	Cedar bevel lap siding

Chickens and Other Creatures

HOW TO BUILD A TOP-BAR BEEHIVE

1. Lay out cutting lines for the insert panels on a piece of 1 × 12 cedar stock. The trapezoid-shaped panels (sometimes called followers) are meant to slide back and forth within the hive cavity, much like a file folder divider. This allows the beekeeper to subdivide the hive space as the honeycombs accumulate. The shape should be 15" wide along the top and 5" wide along the bottom (See diagram, page 31).

2. Cut the insert panels to size and shape and then attach a top cap to the top edge of each panel. The 1 × 2 caps, installed with the flat surface down, should overhang the panels by 1" at each end. Use exterior-rated wood glue and 2" deck screws driven through pilot holes to attach the tops. Also cut 20 top bars from the same 1 × 2s. Use a router or table saw to cut a ¼ × ¼" groove in the bottom of each top bar (inset). The bees use these grooves to create purchase for their hanging honeycombs.

3. Secure the two insert panels upside down on a flat work surface and use them to register the side panels so you can trace the panel locations onto the end panels. Center the end panels against the ends of the side panels, making sure the overhang is equal on each side. Outline the side panel locations, remove the end panels, and drill pilot holes in the outlined area.

4. Attach the end panels to the side panels with glue and 2½" deck screws driven through the pilot holes in the end panels.

5

Cut the parts for the frame that fits around the top of the hive box and fasten them with glue and 1¼" deck screws. The top of the frame should be slightly more than ¾" above the tops of the side panels to provide clearance for the top bars, which will rest on the side panel edges.

6

Attach the legs. First, cut 36"-long legs from 1 × 4 stock and place them over the box ends as shown in the diagram on page 31. Mark cutting lines where the leg tops intersect with the bottom of the frame. If your hive will be on grass or dirt, leave the bottom ends uncut to create a point that will help stabilize the hive. If your hive will be on a hard surface, cut the ends so they are parallel to the tops and will rest flush on the ground. Attach the legs with two or three ⅜ × 2" galvanized lag bolts fitted with washers and nuts.

7

Drill entrance holes and attach the box bottom. On one side panel, drill three 1"-dia. bee entrance holes 2" up from the bottom of the hive. One hole should be centered end to end and the others located 3" away from the center. On the other side, drill a 1"-dia. hole 2" up from the bottom of the hive and 5" from each end. Attach a steel or plastic mesh bottom with roofing nails or narrow crown staples.

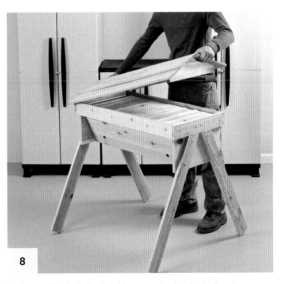

8

Make and install the lid. You can design just about any type of covering you like. Here, a frame with a gable peak is made from cedar stock and then capped with beveled-lap siding (also cedar). The overlap area where the siding fits along the peak ridge should be sealed with clear exterior caulk. Add the inserts and top bars and then fit the lid frame around the box top frame.

Chickens and Other Creatures

SECTION 2

Nourishing Your Garden

In many ways, the self-sufficient home revolves around the garden. Growing and harvesting your own produce enables you to declare some measure of independence from the corporate food chain. It also allows you to control what goes into what you grow, meaning that you'll eat less pesticides, chemicals from processed fertilizer, and other contaminants. And don't forget the money you'll save by raising your own organic produce.

But self-sufficient gardening is something more as well. It's the chance to get involved with your food and truly "get back to the land." There is nothing quite like a few sore muscles and the knowledge that at the end of a day of gardening, you've done simple, good work. You soak up sunshine, get the most wholesome form of exercise you can get, and achieve something tangible and positive. You could hardly ask for a more rewarding outdoor activity.

Of course, gardening involves much more than simply digging in the dirt and dropping a seed in a hole. In fact, it starts with that dirt. Healthy soil will be the foundation of all you do in the garden, and without it, your labor is likely to be much less productive. Much of the fertilizer sold at home centers is like junk food for plants—a quick, sugary rush that leaves your soil more depleted afterward. The best way to build up your soil and make it more productive is with compost, which provides rich, long-lasting nourishment for your soil, which in turn nourishes your plants in the best way possible. You can buy bags of compost, but you can also make your own by recycling your leaves, grass clippings, kitchen waste, coffee grounds, chicken bedding, and more. Just mix them up a compost bin and in a month you'll have rich, nutritious soil to spread around the garden.

Rainwater is another great resource for the garden and lawn, but too many homeowners let rainwater drain away or accumulate where it's not needed (such as around the foundation), and end up relying on their tap water to keep plants green. In this section we'll show you some easy ways to capture your rainwater and use it, instead of wasting it or just letting it run down the storm sewer. You'll also learn about sub-irrigated planters, which are the best way to deliver a steady supply of water to your plants if you're using containers or raised beds.

05 | Compost Bin

The byproducts of yard maintenance and food preparation accumulate rapidly. Everyday yard-care alone creates great heaps of grass clippings, trimmings, leaves, branches, and weeds. Add to this the potato peelings, coffee grounds, apple cores, and a host of organic kitchen leavings. The result is a large mass of organic matter that is far too valuable a resource to be simply dumped into the solid waste stream via curbside garbage collection. Yard waste and kitchen scraps can be recycled into compost and incorporated back into planters or garden beds as a nutrient-rich (and cost-free) soil amendment.

Compost is nature's own potting soil, and it effectively increases soil porosity, improves fertility, and stimulates healthy root development. Besides, making your own soil amendment through composting is much less expensive than buying commercial materials.

So how does garbage turn into plant food? The process works like this: organisms such as bacteria, fungi, worms, and insects convert compost materials into humus, a loamy, nutrient-rich soil. Humus is the end goal of composting. Its production can take a couple of years if left undisturbed, or it can be sped up with some help from your pitchfork and a little livestock manure.

Although composting occurs throughout nature anywhere some organic matter hits the earth, in our yards and gardens it is always a good idea to contain the activity in a designated area, like a compost bin. Functionally, there are two basic kinds of bins: multi-compartment compost factories that require a fair amount of attention from you and hidden heaps where organic matter is discarded to rot at its own pace. Both approaches are valid and both will produce usable compost. The compost bin project shown on page 40 is an example of the more passive style. At roughly 1 cubic yard in volume, it can handle most of your household organic waste and some garden waste. If you have a higher volume of organic waste, you may want to use a two- or three-bin approach (see page 44), which allows you to have piles in different stages of decomposition.

Composting turns yard waste and kitchen scraps into a valuable soil amendment.

Nourishing Your Garden

Called "black gold" by home gardeners, compost can be generated on-site and added to any planting bed, lawn, or container for a multitude of benefits. Sifting the compost before you introduce it to your yard or garden is recommended.

Compost Variables

Air: The best microbes for decomposing plant materials are aerobic, meaning they need oxygen. Without air, aerobic microbes die and their anaerobic cousins take over. Anaerobic microbes thrive without oxygen and decompose materials by putrefaction, which is smelly and slow. Your goal is aerobic activity, which smells musty and loamy, like wet leaves. Improve air circulation in your compost bin by ensuring air passageways are never blocked. Intersperse layers of heavier ingredients (grass clippings, wet leaves) with lighter materials like straw, and turn the pile periodically with a garden fork or pitchfork to promote air circulation.

Water: Compost should be as wet as a wrung-out sponge. A pile that's too wet chokes out necessary air. A too-dry pile will compost too slowly. When adding water to a compost pile, wet in layers, first spraying the pile with a hose, then adding a layer of materials.

Temperature: A fast-composting pile produces quite a bit of heat. On a cool morning, you might notice steam rising from the pile. This is a good sign. Track the temperature of your pile and you'll know how well it's progressing. Aim for a constant temperature between 140 and 150 degrees Fahrenheit, not to exceed 160 degrees Fahrenheit. To warm up a cool pile, agitate it to increase air circulation and add nitrogen-dense materials like kitchen waste or grass clippings. A pile about 3 feet high and wide will insulate the middle of the pile and prevent heat from escaping. You'll know the compost process is complete when the pile looks like dirt and is no longer generating extraordinary heat.

WHAT TO COMPOST, WHAT NOT TO COMPOST

Vegetable plants soak up the materials that make up your compost, and these materials will play a vital role in the development of the vegetables that will grace your dinner table! When in doubt as to what should or shouldn't go into your compost pile for your garden, follow these general guidelines.

GREAT GARDEN COMPOST	NOT FOR COMPOST, PLEASE
"Clean" food scraps—including crushed eggshells, waste, corncobs, vegetable scraps, oatmeal, stale bread, etc.	Fatty or greasy food scraps—including meat bones, grease, dairy products, cooking oils, dressings sandwich spreads, etc.
Vegetable and fruit peelings and leftovers	Fruit pits and seeds—these don't break down well and can attract rodents.
Coffee grounds and filters, tea leaves and tea bags	Metal. Remove the tea bag staples before composting!
Old potting soil	Diseased plant material
Lawn clippings	Weeds—these will only sprout in your garden! Kill the weed seeds and salvage the compostable bits by baking or microwaving the plants before adding them to your compost bin.
Prunings from your yard, chopped up in small pieces	Big chunks of yard debris or plants that are diseased or full of insect pests
Shredded leaves and pine needles	Any plant debris that has been treated with weed killer or pesticides
Shredded newspaper and telephone books—black and white pages only	Glossy color ads or wax-coated book covers
White or brown paper towels and napkins	Colored paper towels and napkins
Wood ash—use sparingly	Coal ash
Cardboard	Pizza boxes or other wax-coated food boxes
Livestock manure	Cat, dog, or other pet waste, which may contain meat products or parasites
Sawdust, wood chips, and woody brush	Sawdust from wood treated with preservatives
Straw or hay—the greener, the better!	
Wilted floral bouquets	

BUILDING A COMPOST BIN

TOOLS & MATERIALS

½" galvanized hardware cloth 36" by 12'
U-nails (fence staples)
2 pairs 2 × 2" butt hinges
2½", 3" deck screws
Pipe or bar crimps
Exterior wood glue
Galvanized finish nails
Exterior wood sealant
Table saw or circular saw
Eye and ear protection
Work gloves

A compost bin can be very plain, or it can have just enough decorative appeal to improve the appearance of a utility area.

CUTTING LIST

KEY	PART	NO.	DIM.	MATERIAL
A	Post	8	1½ × 1¾ × 48"	Cedar
B	Door rail	2	1½ × 3½ × 16"	Cedar
C	Door rail	2	1½ × 1¾ × 16"	Cedar
D	Door stile	4	1½ × 1¾ × 30½"	Cedar
E	Panel rail	3	1½ × 3½ × 32½"	Cedar
F	Panel rail	3	1½ × 1¾ × 32½"	Cedar
G	Panel stile	3	1½ × 3½ × 30½"	Cedar
H	Infill	16	¾ × 1½ × 30½"	Cedar
I	Filler	80	¾ × 1½ × 4"	Cedar
J	Panel grid	12	¾ × 1½" × Cut to fit	Cedar
K	Grid frame-v	16	¾ × 1½" × Cut to fit	Cedar
L	Door frame-h	4	¾ × 1½" × Cut to fit	Cedar
M	Top rail-side	2	1½ × 1¾ × 39"	Cedar
N	Top rail-back	1	1½ × 1¾ × 32½"	Cedar
O	Front spreader	1	1½ × 3½ × 32½"	Cedar

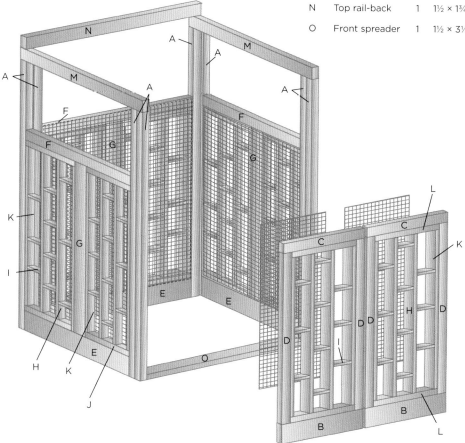

The Beginners Guide to Self Sufficency Projects for the Home

HOW TO BUILD A COMPOST BIN

Prepare the wood stock. At most building centers and lumber yards, you can buy cedar sanded on all four sides, or with one face left rough. The dimensions in this project are sanded on all four sides. Prepare the wood by ripping some of the stock into 1¾"-wide strips (do this by ripping 2 × 4s down the middle on a tablesaw or with a circular saw and cutting guide).

Cut the parts to length with a power miter saw or a circular saw. For uniform results, set up a stop block and cut all similar parts at once.

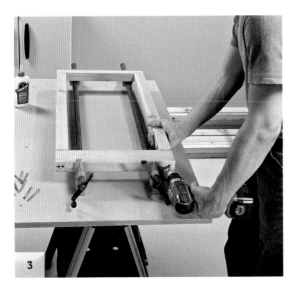

Assemble the door frames. Apply exterior-rated wood glue to the mating parts and clamp them together with pipe or bar clamps. Reinforce the top joints with 3" countersunk deck screws (two per joint). Reinforce the bottom joints by drilling a pair of ⅜"-dia. × 1"-deep clearance holes up through the bottom edges of the bottom rails and then driving 3" deck screws through these holes up into the stiles.

Assemble the side and back panels. Clamp and glue the posts and rails for each frame, making sure the joints are square. Then, reinforce the joints with countersunk 3" deck screws.

(continued)

HOW TO BUILD A COMPOST BIN (continued)

Hang the door frames. With the posts cut to length and oriented correctly, attach a door frame to each post with a pair of galvanized butt hinges. The bottoms of the door frames should be slightly higher than the bottoms of the posts. Temporarily tack a 1 × 4 brace across both door bottom rails to keep the doors from swinging during construction.

Join the panels and the door assembly by gluing and clamping the parts together and then driving 2½" countersunk deck screws to reinforce the joints. To stabilize the assembly, fasten the 2 × 4 front spreader between the front, bottom edges of the side panels. Make sure the spreader will not interfere with door operation.

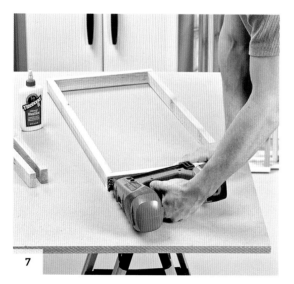

Make the grids for the panel infill areas. Use 1 × 2 cedar to make all parts. Use exterior glue and galvanized finish nails to connect the horizontal filler strips to the vertical infill pieces. Vary the heights and spacing of the filler for visual interest and to make the ends accessible for nailing.

Frame the grids with 1 × 2 strips cut to the correct length so each frame fits neatly inside a panel or door opening. Install the grid frames in the openings, making sure all front edges are flush.

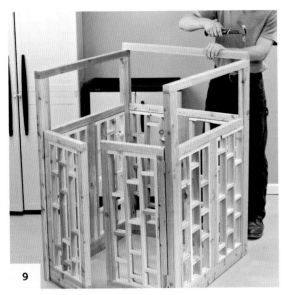

Attach the top rails that conceal the post tops and help tie the panels together. Attach the sides first using exterior glue and galvanized finish nails. Then, install the back rail on top of the side rails. Leave the front of the project open on top so you can load, unload, and turn over compost more easily.

Line the interior surfaces of the compost bin with ½" galvanized hardware cloth. Cut the hardware cloth to fit and fasten it with fence staples, galvanized U-nails, or narrow-crown pneumatic staples (⅝" minimum) driven every 6" or so. Make sure you don't leave any sharp edges protruding. Grind them down with a rotary tool or a file.

Set up the bin in your location. It should not be in direct contact with any structure. If you wish, apply a coat of exterior wood sealant to all wood surfaces—use a product that contains a UV inhibitor. **TIP:** Before setting up your compost bin, dig a hole just inside the area where the bin will be placed. This will increase you bin's capacity.

A fast-burning compost pile requires a healthy balance of "browns" and "greens." Browns are high in carbon, which food energy microorganisms depend on to decompose the pile. Greens are high in nitrogen, which is a protein source for the multiplying microbes.

Nourishing Your Garden

06 | Two-Bin Composter

When it's time to get serious about composting, a multiple-bin system is the way to go. They're designed to produce a large volume of compost in a short time. The idea is to develop a nice, big heap in one bin, then start turning it over by shoveling it into the neighboring bin, then back to the first bin, and so on. Turning greatly speeds decomposition (plus, it gives you a little exercise in the process). Depending on the compost materials, turning is recommended every 5 to 10 days. A two-bin composter lets you flip the heap back and forth between bins until the compost is ready, then you can store it one bin and use the other bin to start building the next heap.

This design facilitates turning with its removable divider between the bins. Simply slide the divider up and out of the way for easy shoveling. The front sides of the bins are full-width gates, providing easy access to the bins for moving material in or out.

But perhaps the best feature of this composter has nothing to do with production; it's all about appearances. As much as self-sufficient homeowners and gardeners love the idea of composting, few can honestly say they like the look of a compost heap. (And the aesthetics of plastic barrels or trashcan composters need no further criticism.) You may not see a lot of bin-type composters with cedar pickets, decorative posts, and traditional gates, but what would you rather look at: a pile of rotting garbage or a well-built picket fence?

THREE-PEAT ENOUGH

A three-bin system uses the same idea as a two-bin composter, but the additional bin helps make the process even more continuous. Once your heap is ready, flip it into bin two and use bins two and three for turning. This leaves bin one open for compiling the next heap. You can easily adapt this two-bin composter design to create a three-bin version. Just extend the overall length by a third, and create another center divider and gate. The two stringers along the backside of the structure can be cut from 12' 2 × 4s.

Composting is key in a truly self-sufficient garden, and the only thing better than an active composting bin, is a doubly active bin.

BUILDING A TWO-BIN COMPOSTER

CUTTING LIST

KEY	NO.	PART	DIMENSION	MATERIAL*
A	6	Posts	3½ × 3½ × 60"	4 × 4 PT pine
B	49	Pickets	¾ × 3½ × 36"	1 × 4 PT pine
C	2	Rear rail	1½ × 3½ × 93"	2 × 4 PT pine
D	4	Side rail	1½ × 3½ × 45"	2 × 4 PT pine
E	4	Hinge blocks	1½ × 3½ × 5"	2 × 4 PT pine
F	4	Gate rails	1½ × 3½ × 39¼"	2 × 4 PT pine
G	2	Latch block	1½ × 3½ × 3½"	2 × 4 PT pine
H	4	Divider panel stops	1½ × 1½ × 30"	2 × 2 PT pine
J	2	Divider rail	1½ × 3½ × 34"	2 × 4 PT pine

*Use pressure-treated lumber rated for ground contact or all-heart cedar, redwood, or other naturally rot-resistant species.

TOOLS & MATERIALS

4' level
Posthole digger
Miter saw
Cordless drill and bits
Gravel
Deck screws 1⅝", 3½"
Exterior-grade construction adhesive
½ × ½" galvanized hardware cloth with staples (optional)
(4) 3½ gate hinges with screws
(2) Gate latches with screws
4 × 4 post caps (optional)
Chalk
Framing square
Eye and ear protection
Work gloves

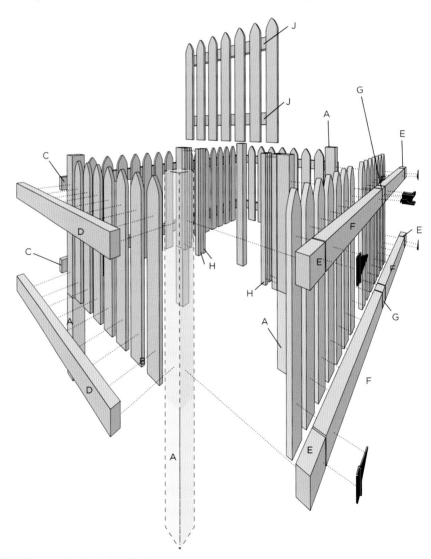

The Beginners Guide to Self Sufficiency Projects for the Home

HOW TO BUILD A TWO-BIN COMPOSTER

1

Choose a flat, level site at least 6 × 10' that allows for movement in front of the bins. Cut the back rails and posts to length, then lay them out on a flat surface. Square the assembly by making sure the diagonals are equal. Fasten the rails to the posts with 3½" deck screws.

2

Set the assembled back wall in place, then mark the hole locations with chalk or a shovel. Dig holes 8" in diameter by 24" deep at each post location.

3

Plumb and level the posts using wood braces. Fill the holes with gravel and dirt. The bottom rails should be roughly 6" to 8" above ground level.

(continued)

Nourishing Your Garden

BUILDING A TWO-BIN COMPOSTER (continued)

Dig holes for the front posts, using a framing square and a side rail (or the 3-4-5 method) to locate the correct position. Put the posts in and fasten the top rail to both front and rear posts to help hold the front post plumb and level. Fill the holes with alternating layers of gravel and dirt and attach the bottom rails. Position the center post so that it's the same distance from the corner posts and in line with both of them.

At the front corner posts and front center post, measure and cut six short 2 × 4s to cover the front faces of the posts (5" long at the corners and 3½" at the center). These continue the runs of the stringers and will serve as mounting blocks for the gate hinges and latches. Install the blocks as you did with the rear stringers, but predrill and countersink all screw holes to avoid cracking the short pieces of wood.

Install the pickets along the back and sides with 1⅝" deck screws. Keep the pickets 1½" to 2" above the ground to prevent rot. The fastest way to install the pickets is to make spacer blocks. For this design, the seven pickets on the sides were spaced 1⅝" apart and the back and front pickets were spaced 1⅞" apart. To find the spacing for a different size bin, just subtract the total width of seven pickets (or however many you use) from the distance between the posts, then divide the result by 8 (the number of spaces between pickets). Use a 5"-high block of wood at the top to quickly establish the height for each picket.

Begin constructing the gates. Set the rails on your work surface so they are parallel and spaced the same distance apart as the bin stringers. Space the pickets 1⅞" apart like the back wall, but start from the center—line up the center of the first picket with the center of the rail, then work to each side so you end up with roughly 1⅞" between the last picket and the post. Fasten the pickets to the rails with 1⅝" screws and construction adhesive. The construction adhesive helps prevent the gate from sagging over time. Check the assembly for square as you work.

8 Clamp or screw a straight piece of wood across the three front posts to support the gates while the hinges and latches are attached. Hang the gates using gate hinges—both gates open out and away from the center post. Install latch hardware for each gate so it locks closed at the center post.

9 Cut the four stops for the sliding divider panel from 2 × 2s. Make marks at 1¼" in from the edges on both posts. Install the 2 × 2 stops against these lines with predrilled 3½" screws, creating a 1" slot at the center. The outside edges of the stops will overhang the edges of the posts by about ¼".

10 Build the divider panel using the same construction techniques used for the gates, but extend the outside pickets beyond the 2 × 4 rails so they fit into the channels between the stops. Space the pickets 2⅛" apart. The total width of the panel should be about ½" narrower than the distance between the posts so that the panel can slide in and out without binding. To keep the divider panel at the same height as the rest of the enclosure, screw in a small wood stop or a few screws near the bottom of the slot.

UPGRADES

Add a touch of beauty to your compost bin by attaching a decorative post cap to the top of each 4 × 4 post. You can find a wide variety at home centers or online. You also may want to fasten a 2'- wide strip of ½" hardware cloth down to the ground around the inside of the composter to help keep all the compost inside the bin.

07 | Collecting Rainwater

Practically everything around your house that requires water loves the natural goodness that's provided with soft rainwater. With a simple rain barrel, you can collect rainwater to irrigate your garden or lawn, water your houseplants, or top off swimming pools and hot tubs. A ready supply of rainwater is also a reliable stand-by for emergency use if your primary water supply is interrupted.

Collecting rainwater runoff in rain barrels can save thousands of gallons of tap water each year. A typical 40 × 40-foot roof is capable of collecting 1,000 gallons of water from only 1 inch of rain. A large rainwater collection system that squeezes every drop from your roof can provide most—or sometimes all—of the water used throughout the home, if it's combined with large cisterns, pumps, and purification processing.

Sprinkling your lawn and garden can consume as much as 40 percent of the total household water use during the growing season. A simple rain barrel system that limits collected water to outdoor (nonpotable) use only, like the rain barrels described on the following pages, can have a big impact on the self-sufficiency of your home, helping you save on utility expenses and reducing the energy used to process and purify water for your lawn and garden. Some communities now offer subsidies for rain barrel use, offering free or reduced-price barrels and downspout connection kits. Check with your local water authority for more information. Get smart with your water usage, and take advantage of the abundant supply from above.

NOTE: The collection of rainwater is restricted in some areas. Check with your municipality if you are unsure.

Rainwater that is collected in a rain barrel is free of the chemical additives and minerals usually found in tap water. This soft, warm (and free) water is perfect for gardens, lawns, and indoor plants like orchids that don't do well with tap water.

Rain Barrels

Rain barrels, either built from scratch or purchased as a kit, are a great way to irrigate a lawn or garden without running up your utilities bill. The most common systems include one or more rain barrels (40 to 80 gallons) positioned below gutter downspouts to collect water runoff from the roof. A hose or drip irrigation line can be connected to spigot valves at the bottom of the rain barrel. You can use a single barrel, or connect several rain barrels in series to collect and dispense even more rainwater.

Plastic rain barrel kits are available for purchase at many home centers for around $100. If kit prices aren't for you, a rain barrel is easy to make yourself for a fraction of the price. The most important component to your homemade barrel is the drum you choose.

Obtaining a Barrel

Practically any large waterproof container can be used to make a rain barrel. One easily obtained candidate is a trash can, preferably plastic, with a snap-on lid. A standard 32-gallon can will work for a rain barrel, but if you can find a 44-gallon can, choose it instead. Although wood barrels are becoming more scarce, you can still get them from wineries. A used 55-gallon barrel can be obtained free or for a small charge from a bulk food supplier. Most 55-gallon barrels today are plastic, but some metal barrels are still floating around. Whatever the material, make sure the barrel did not contain any chemical or compound that could be harmful to plants, animals, or humans. If you don't know what was in it, don't use it. Choose a barrel made out of opaque material that lets as little light through as possible, reducing the risk of algae growth.

A barrelful of water is an appealing breeding ground for mosquitoes and a perfect incubator for algae. Filters and screens over the barrel opening should prevent insect infestation, but for added protection against mosquitoes add one tablespoon of vegetable oil to the water in the barrel. This coats the top surface of the stored water and deprives the larvae of oxygen.

TOOLS & MATERIALS
Barrel or trash can
Drill with spade bit
Jigsaw
Hole saw
Barb fitting with nut for overflow hose
1½" sump drain hose for overflow
¾" hose bibb or sillcock
¾" male pipe coupling
¾" bushing or bulkhead connector
Channel-type pliers
Fiberglass window screening
Cargo strap with ratchet
Teflon tape
Silicone caulk

HOW TO MAKE A RAIN BARREL

Cut a large opening in the barrel top or lid. Mark the size and shape of your opening—if using a bulk food barrel, mark a large semicircle in the top of the barrel. If using a plastic garbage can with a lid, mark a 12"-dia. circle in the center of the lid. Drill a starter hole, and then cut out the shape with a jigsaw.

Install the overflow hose. Drill a hole near the top of the barrel for the overflow fitting. Thread the barb fitting into the hole and secure it to the barrel on the inside with the retainer nut and rubber washer (if provided). Slide the overflow hose into the barbed end of the barb elbow until the end of the hose seats against the elbow flange.

Drill the access hole for the spigot (either a hose bibb or sillcock, brass or PVC). Tighten the stem of the sillcock onto a threaded coupling inserted into the access hole. Inside the barrel, a rubber washer is slipped onto the coupling end and then a threaded bushing is tightened over the coupling to create a seal. Apply a strip of Teflon tape to all threaded parts before making each connection. Caulk around the spigot with clear silicone caulk.

Screen over the opening in the top of the barrel. Lay a piece of fiberglass insect mesh over the top of the trash can and secure it around the rim with a cargo strap or bungee cord that can be drawn drum-tight. Snap the trash can lid over the top. Once you have installed the rain barrel, periodically remove and clean the mesh.

Installing a Rain Barrel

Whether you purchase a rain barrel or make your own from scratch or a kit, how well it meets your needs will depend on where you put it and how it is set up. Some rain barrels are temporary holding tanks that store water runoff just long enough to direct it into your yard through a hose and drip irrigation head. Other rain barrels are more of a reservoir that supplies water on-demand by filling up watering cans or buckets. If you plan to use the spigot as the primary means for dispensing water, you'll want to position the rain barrel well off the ground for easy access (raising your rain barrel has no effect on water pressure).

In addition to height, other issues surrounding the placement of your rain barrel (or rain barrels) include the need to provide a good base, orientation of the spigot and overflow, the position relative to your downspouts, and how to link more than one rain barrel together. **NOTE:** If you live in a cold climate, it's a good idea to drain the rain barrel and close it for the season to avoid having it crack when the water freezes.

TOOLS & MATERIALS
Drill/driver
Screwdriver
Hacksaw
Rainbarrel
Hose & fittings
Base material (pavers)
Downspout adapter and extension
Teflon tape
Sheet metal screws
Eye and ear protection
Work gloves

HOW TO INSTALL A RAIN BARREL

Select a location for the barrel under a downspout. Locate your barrel as close to the area you want to irrigate as possible. Make sure the barrel has a stable, level base.

Install the spigot. Some kits may include a second spigot for filling watering cans. Use Teflon tape at all threaded fittings to ensure a tight seal. Connect the overflow tube, and make sure it is pointed away from the foundation.

Cut the downspout to length with a hacksaw. Reconnect the elbow fitting to the downspout using sheet-metal screws. Attach the cover to the top of the rain barrel. Some systems include a cover with porous wire mesh, to which the downspout delivers water. Others include a cover with a sealed connection (next step).

Link the downspout elbow to the rain barrel with a length of flexible downspout extension attached to the elbow and the barrel cover.

VARIATION: If your barrel comes with a downspout adapter, cut away a segment of downspout and insert the adapter so it diverts water into the barrel.

Connect a drip irrigation tube or garden hose to the spigot. A Y-fitting, like the one shown here, will let you feed the drip irrigation system through a garden hose when the rain barrel is empty.

If you want, increase water storage by connecting two or more rain barrels together with a linking kit, available from many kit suppliers.

Nourishing Your Garden

08 | Channeling Rainwater

An arroyo is a dry streambed or watercourse in an arid climate that directs water runoff on the rare occasions when there is a downfall. In a home landscape an arroyo may be used for purely decorative purposes, with the placement of stones evoking water where the real thing is scarce.

Or it may serve a vital water-management function, directing storm runoff away from building foundations to areas where it may percolate into the ground and irrigate plants, creating a great spot for a rain garden. This water management function is becoming more important as municipalities struggle with an overload of storm sewer water, which can degrade water quality in rivers and lakes. Some communities now offer tax incentives to homeowners who keep water out of the street.

When designing your dry streambed, keep it natural and practical. Use local stone that's arranged as it would be found in a natural stream. Take a field trip to an area containing natural streams and make some observations. Note how quickly the water depth drops at the outside of bends where only larger stones can withstand the current. By the same token, note how gradually the water level drops at the inside of broad bends where water movement is slow. Place smaller river-rock gravel here, as it would accumulate in a natural stream.

Large heavy stones with flat tops may serve as step stones, allowing paths to cross or even follow dry stream beds.

The most important design standard with dry streambeds is to avoid regularity. Stones are never spaced evenly in nature and nor should they be in your arroyo. If you dig a bed with consistent width, it will look like a canal or a drainage ditch, not a stream. And consider other yard elements and furnishings. For example, an arroyo presents a nice opportunity to add a landscape bridge or two to your yard.

An arroyo is a drainage swale lined with rocks that directs runoff water from a point of origin, such as a gutter downspout, to a destination, such as a sewer drain or a rain garden.

HOW TO BUILD AN ARROYO

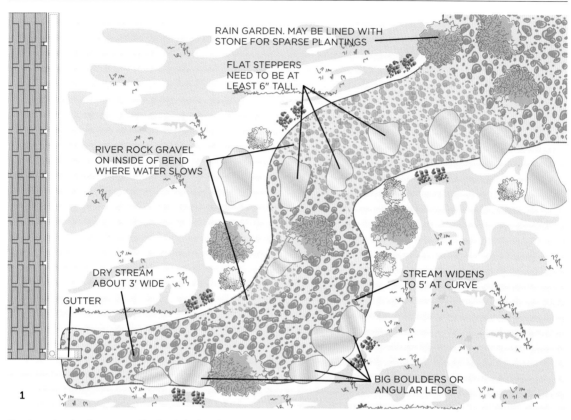

Create a plan for the arroyo. The best designs have a very natural shape and a rock distribution strategy that mimics the look of a stream. Arrange a series of flat steppers at some point to create a bridge.

TOOLS & MATERIALS

Landscape paint
Carpenter's level
Spades
Garden rake
Wheelbarrow
Landscape fabric or 6-mil black plastic
Mulch
8"-thick steppers
6 to 18" dia. river-rock boulders
¾ to 2" river rock
Native grasses or other perennials for banks
Eye and ear protection
Work gloves

Lay out the dry stream bed, following the native topography of your yard as much as possible. Mark the borders and then step back and review it from several perspectives.

Excavate the soil to a depth of at least 12" (30cm) in the arroyo area. Use the soil you dig up to embellish or repair your yard.

Rake and smooth out the soil in the project area. Install an underlayment of landscape fabric over the entire dry stream bed. Keep the fabric loose so you have room to manipulate it later if the need arises.

Place flagstone steppers or boulders with relatively flat surfaces in a stepping-stone pattern to make a pathway across the arroyo (left photo). Alternately, create a "bridge" in an area where you're likely to be walking (right photo).

(continued)

HOW TO BUILD AN ARROYO (continued)

6 Add more stones, including steppers and medium-size landscape boulders. Use smaller aggregate to create the stream bed, filling in and around, but not covering, the larger rocks.

WHAT IS A RAIN GARDEN?

A rain garden is simply a shallow, wide depression at least 10' away from a basement foundation that collects storm water runoff. Rain gardens are planted with native flood-tolerant plants and typically hold water for only hours after rainfall. Check your local garden center or extension service to find details about creating rain gardens in your area.

7 Dress up your new arroyo by planting native grasses and perennials around its banks.

ALTERNATIVE: CREATE A SWALE

A swale is basically just a gently sloping trench filled with grass or plants instead of stone. Like an arroyo, a swale channels runoff from a gutter or low area to a rain garden or other section of your yard. Use stakes or spray paint to mark a swale route that directs water away from the problem area toward a runoff zone.

Remove sod and soil from the marked zone. Cut the sod carefully (or rent a sod cutter) and set it aside to reuse if it's in good condition.

Level the trench by laying a 2 × 4 board with a carpenter's level on the foundation. Distribute soil so the base is even, and so that it slopes enough to move the water in the direction you want it to go.

Lay sod in the trench to complete the swale. Compress the sod and water the area thoroughly to check drainage.

Nourishing Your Garden

SECTION 3

Garden Projects

Most of us don't have acreage to plant; instead, we're working with relatively small urban and suburban lots where planting space is limited, where vegetables and flower gardens have to share space with kids, yard furniture, decks, barbecues, shade trees, and pets. But limited space doesn't mean you have to limit your ambitions. The trick is to be creative with the space you have, and to grow more with less.

In this section we'll show you how to expand your growing space and the size of your crop by growing more efficiently and by growing upward—where the space is virtually unlimited. Training plants to grow on trellises like the ones shown in this section will substantially increase yields of tomatoes, beans, peppers, and other vegetables by giving them more room and light and by keeping them off the ground and away from mold and insects. Cold frames increase your yield by extending the growing season, allowing you to start planting in late winter or early spring. Raised beds let you transform ordinary, tired dirt into state-of-the-art soil. Instead of trying to work infertile, rocky dirt, you just build a box on top of it and fill it with a rich, high-quality growing medium designed to produce high yields.

09 | Starting and Transplanting Seedlings

Add weeks to your garden's growing season by starting seeds indoors, then transplanting them to the garden after the danger of frost is past. Seedlings are available for purchase at the garden center in the spring, of course, but starting your own at home presents a number of advantages:

- Buying seeds is less expensive than buying seedlings.
- You can cull out all but the strongest seedlings, which will hopefully result in stronger plants and a more bountiful crop.
- Garden centers sell seeds for a diverse and varied array of plants but seedlings for only the most common species. Seed catalogs introduce an entirely new selection as well.
- You can be certain that unwanted pesticides have not been used on the plants in your garden, even in their infancy.

Start your seeds 8 to 10 weeks before you plan to transplant them into your garden. To get started, you'll need a few small containers, a suitable growing medium, and a bright spot for the seedlings to grow—either a sunny window that receives at least six hours of bright sunlight per day, a greenhouse, or a planting table in your home that's illuminated by artificial grow lights. If you're planning to raise your seedlings by artificial light, position one or two fluorescent lighting fixtures fitted with 40-watt, full-spectrum bulbs about 6 inches above the seedlings. Leave fluorescent lights on for 12 to 16 hours a day—many gardeners find it helpful to connect the fixture to a timer to ensure their plants receive adequate light each day.

Using colored cups as starter containers has the advantage of letting you color-code your plants so you can avoid any confusion. Beginning gardeners often have trouble distinguishing one pot from another when they are still seedlings. Here, the red cups clearly communicate "tomato."

Seedlings need a lot of water, sunlight, and warmth in their infancy. A kitchen window or greenhouse is an ideal growing environment for propagating plants.

Garden Projects

A 4', two-bulb fluorescent light fixture that can be raised or lowered over a table is really all you need to start your vegetable plants indoors.

GROWING MEDIUMS

If you plan to use your own garden soil or compost, prepare your seedling containers in the fall, before the ground gets too cold and wet.

Almost any small container can be used to grow seedlings. Just make sure the container you use is clean and hasn't had contact with any chemical that could be poisonous to plants. Also, remember to cut a drainage hole in the bottom of your container before filling it with soil. Drainage is very important to ensure that your plants are well ventilated. Excessively moist soil can result in mold or other diseases, as well. Good options for seedling containers include:

- Peat pots or pellets
- Cans (any size)
- Used yogurt cups

- Fiber cubes
- Used plastic tubs (i.e., sour cream, cottage cheese, or margarine containers
- Egg cartons
- Used plastic jugs
- Small paper cups

Starting Seedlings

1. **Planting:** Sow three to four seeds in each container according to the instructions on the seed packet—as a general rule, large seeds should be buried and small seeds can be sprinkled on top of the soil. Label the container with the type of plant and the date your seeds were planted.

2. **Germination:** Water the seeds whenever the containers look dry. Until the seeds sprout, keep seedlings in a dark, warm space. Cover germinating seeds with plastic bags or plastic wrap. Open the plastic for a few hours every few days to let the soil breathe, then re-close.

3. **Sprouts:** When the seeds sprout, remove plastic covering and move them into direct light. Seedlings need lots of light to grow. Keep the soil medium moist but not soggy. Remember, multiple light waterings are better for seedlings than the occasional soaking.

4. **Culling:** When the true leaves appear (see illustration), cut off all but the strongest seedling in each container at soil level. Do not pull up the unwanted seedlings, as this may damage the roots of the seedling you're cultivating. You may also choose to fertilize every week or so as your seedlings grow.

Hardening Off & Transplanting

When your seedlings have four to eight true leaves, they should be hardened off, then transplanted to the garden. Hardening off is the process of gradually introducing the plant to outdoor conditions so it is not shocked when you move it outside permanently. About two weeks before planting, place your seedlings outdoors for an hour the first day, and then gradually increase the time until your seedlings spend the whole day outside. Protect seedlings from wind and do not expose them to the midday

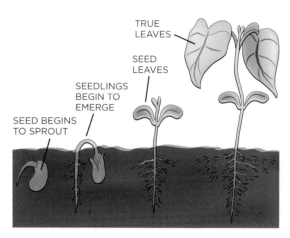

The four stages in the growth of a seedling are illustrated above. Note that seed leaves and true leaves serve different purposes and will look different. When a plant has four to eight true leaves, it is ready for transplanting.

sun for the first few days. Stop fertilizing seedlings during the last week. A cold frame is a great environment for hardening off seedlings. Seedlings can stay in a cold frame for two to three weeks, gradually getting used to the cooler air and chilly nights before they go out into the garden. Open the lid of the cold frame a little bit more each day.

Transplant your seedlings into the garden on a cloudy day or in late afternoon to avoid excessive drying from the sun. Remove seedlings gently from their containers, holding as much soil as possible around the roots (containers that are pressed from peat are not intended to be removed). Place each into a hole in your garden, spreading the roots carefully, then pack soil around the seedling to hold it straight and strong. Thoroughly soak all seedlings with a very gentle water spray after they've been planted. If you have a rain barrel or another source for untreated water, this is a perfect application for it: the chlorine in most municipal water can be harmful to delicate plants.

WHAT TO GROW

Most plants that grow well in your garden will also thrive in containers. Root vegetables are perhaps the only exception to this rule. Keep in mind that the larger the plant, the larger the container you'll need. Generally, the plant should not be more than twice the height of the pot or 1½ times as wide. Use the guidelines below as a rough guide.

Containers 4 to 6" deep: Mustard greens, radishes, and spinach can all grow in shallow containers.

Containers at least 8" deep: Corn (container must be at least 21" wide, however, and house at least three plants to assure pollination), kale, lettuce.

Containers at least 12" deep: Beans, beets, brussel sprouts, cabbage (should also be pretty wide), carrots, chard, collards, kohlrabi, onion, peas, turnips, zucchini.

Containers at least 16" deep: Cucumber, eggplant, peppers.

Containers at least 20" deep: Broccoli, bok choy, Chinese cabbage.

More than 24" deep: Squash, tomatoes.

Spinach, leaf lettuce, and a few shallots coexist in this self-watering planter. Self-watering containers have a water reservoir below to keep the soil in the pot moist. All you have to do is keep the reservoir full, and rainfall may even take care of this for you.

Container Types and Recommendations

As a container gardener, you'll quickly discover that the universe of usable containers is infinitely larger than the plain clay flowerpot. Essentially, any sturdy, watertight container will do. Large containers like wine barrels or old wash tubs and smaller containers like an ice cream pail or 5-gallon bucket can all be good for different kinds of plants. Large wooden troughs and DIY planter boxes can be customized to your garden (and are fun to make too). When building your own planters, it's a good idea to line the inside with landscape fabric before adding potting soil to protect the wood from rot and to make it easier to empty out soil after a season.

Always make sure that the container you choose did not previously hold any kind of chemical and, if it does not already have them, drill drainage holes near the bottom of the container before filling with soil. If you'll be using large containers, it's usually a good idea to place them on a platform fitted with casters before filling them with potting soil.

Self-watering pots make container gardening less of a drain on your time. These containers are, essentially, a flowerpot set just above a reservoir of water. With this type of container, the soil above will wick up moisture from the reservoir as it needs it—keeping the soil consistently moist throughout and eliminating the possibility of over-watering. With a self-watering container, you may only need to add water every three to four days, and your plants will likely be less stressed. Ideally, your plants will therefore provide a more sizable crop at the end of the season.

Many varieties of vegetables and fruits can be grown in pots, planters, and other vessels—just make sure to select a container that is the right size for your plant and always add ample drainage holes if they're not already present.

10 | Clothesline Trellis

Modifying or repurposing a clothesline support to serve as a trellis is not a new idea, but it's certainly a good one. It's also kind of a head-slapper, as in, "Why didn't I think of that?" After all, you've got this tall, sturdy, utilitarian structure taking up space in a sunny spot that's easy to reach from the house . . . so why not grow some plants on it?

If you don't already have a clothesline support or two that you can turn into a trellis, you can build this one from scratch. The construction is easier than it looks. All of the beams and uprights are joined with special timber screws, so there's no complex or custom-fit joinery. And you can build the entire trellis in your shop or garage, then dig a couple of holes and get it set up in one go.

The basic structure of the trellis is inspired by the Torii, a traditional Japanese gateway to a shrine or other sacred place. The overhanging top beam, or lintel, is a characteristic feature for this type of structure, and in this case can be used to support hanging plants or wind chimes or simply be left as is for a clean look. The vertical spindles in the center of the trellis are made with 1½-inch-square pressure-treated stock. (You can also use cedar or redwood.) They're offset from one another in an alternating pattern for a subtle decorative effect. You can change the spacing of the spindles as needed to suit your plants, or even use a different material, such as round spindles, wire, or string.

This trellis makes a great garden feature that looks good year-round and can serve as a focal point or a divider between landscape zones. You can build just one trellis and run the clotheslines between the trellis and a fence, your house or garage, or a garden shed or other outbuilding.

A trellis such as this is not only very attractive all in its own right, it also serves two functional roles—holding up plants and holding up laundry!

BUILDING A CLOTHESLINE TRELLIS

CUTTING LIST (ONE TRELLIS)

KEY	NO.	PART	DIMENSION	MATERIAL*
A	2	Post	3½ × 3½" × 10'	4 × 4
B	1	Lintel	3½ × 3½ × 81"	4 × 4
C	2	Crossbeam	3½ × 3½ × 47"	4 × 4
D	7	Spindle	1½" × 1½ × 47½"	2 × 2
E	1	Spreader	3½ × 3½ × 8½"	4 × 4

*All lumber can be pressure-treated or all-heart cedar or redwood or other naturally rot-resistant wood.

TOOLS & MATERIALS

Miter saw
Cordless drill and bits
Nail set
Tongue-and-groove pliers or adjustable wrench
Posthole digger
6" self-drilling timber screws (24)
2" exterior finish nails
Optional: ⅜ × 2¾" galvanized or stainless-steel screw eyes or screw hooks (4, with lag-screw threads)
Gravel
Level
Eye and ear protection
Work gloves

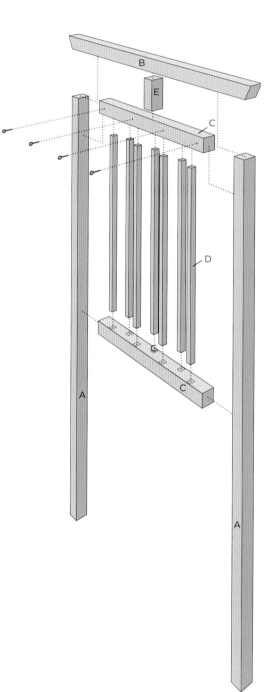

HOW TO BUILD A CLOTHESLINE TRELLIS

Using a miter saw or circular saw, cut both ends of the crossbeams square. Cut the ends of the lintel at 30. Cut the top end of each 10' post to ensure a clean, square cut with no splits; the bottom end will be buried at least 3' deep, so overall length isn't critical.

Mark the inside faces of the two posts for the crossbeam locations, using a square to draw layout lines across the post faces. Mark the underside of the lintel in the same way; it is centered over the long posts, while the center post is centered on the lintel.

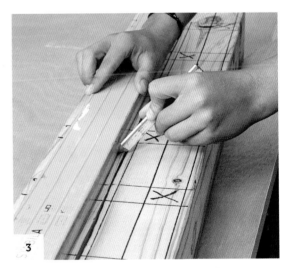

Mark the 2 × 2 spindle locations on the crossbeams. For more contrast, you can offset the locations—but remember to mark the opposing sides of the crossbeam as mirror images. Here the spindles are placed ¼" in from the edges and spaced 4⅝" apart.

(continued)

Garden Projects

HOW TO BUILD A CLOTHESLINE TRELLIS (continued)

4

Test-fit the frame assembly on a flat work surface. Fasten the crossbeams to the posts with two 6" self-drilling timber screws at each joint. You may need to drill pilot holes if the screws are difficult to drive. Drive the screws with either a drill or an impact driver and a hex-type nut driver or other bit (special bits often come with boxes of timber screws). If your 4 × 4s are well-dried, you can attach the lintel now; if not, save some back strain and bolt it on after the posts are upright.

5

Dig two holes at least 3′ deep for the posts. Shovel a few inches of gravel into each hole, then tip the whole assembly in (use a helper—wet treated wood is heavy). Plumb and brace the posts with 1 × 2s.

6

Check that the crossbeams are level. To raise one side, simply add a little gravel under the post. Fill the postholes with alternating layers of soil and gravel. Or you can fill the holes with concrete. Check level and plumb as you fill the holes. Tamp the dirt and gravel so that the posts are firmly locked in place.

Set the lintel in position, aligning it with the marks made earlier, then fasten it in place with timber screws.

Cut the 2 × 2s to fit tightly, then fasten with predrilled 6d casing nails or a pneumatic nailer. Use two nails at each end.

CLOTHESLINES FOR CLIMBERS

Here are a few tips for planning and setting up a decorative and highly functional clothesline system:

- When the leaves are out, the trellis can create a nice a shady spot for a bench on the side opposite the clotheslines.

- A typical load of laundry needs about 35 linear feet of clothesline and weighs about 15 to 18 pounds after spinning in the washing machine. Therefore, running three or four 12' to 15' lines strikes a good balance for minimizing sag on the loaded lines while providing plenty of room for hanging.

- Clothesline materials include solid-metal wire, stranded wire or cable, plastic-coated cable, and traditional clothesline rope. Wire lines last longer and stretch less than rope, but many clothesline devotees prefer rope for its natural look and feel as well as its thickness and texture, which make it ideal for gripping fabric and clothespins.

- Pulleys allow you to hang and retrieve clothes from one standing position. Metal pulleys are strong and won't break down with UV exposure (like plastic wheels will), but make sure all metal parts are rust-resistant (stainless steel is best).

- Turnbuckles provide means for tensioning wire lines without having to restring or reclamp them. Tighten rope lines by simply retying them, or you can add a hook and a trampoline spring to maintain tension and make it easy to remove the line without untying it.

Garden Projects

11 Pallet Planter

Pallets are so abundant in this country that they're often just left by the curb for people to take as firewood. Despite this lack of perceived value, they can actually be quite useful to the thrift-minded self-sufficient homeowner. They can be perfect for a wide variety of recycled projects like compost bins, furniture, fuel and, yes, planters.

Because pallets have to support thousands of pounds, they're generally made from tough hardwoods without big knots. When the wood is cleaned up and sanded, it can look surprisingly attractive. Pallet wood may not last as long outdoors as cedar or treated wood (unless you find a pallet made from white oak), but since it's free you can just replace it when it rots. In the meantime, it looks great.

There are many different ways to turn a pallet into a planter, but don't count on just pulling out all the nails and reusing all the pieces. It can be done, but hardwood grips nails a lot tighter than the softwoods used for construction lumber, and the wood slats often crack before the nail pulls out. (It's wise to always grab a few extra pallets for any project, precisely for this reason.) If you have to remove a few pieces, lever carefully under the wood slats, and use a reciprocating saw outfitted with a metal-cutting blade to cut nails that are hard to pull.

However, the easiest way to use pallets is as is—as we've done in this project. We designed this one to be vertical, but you can also lay it flat and stack a few underneath so that the plants are at a comfortable height, or combine several in a stairstep design.

Look for pallets in industrial and commercial areas. If you're lucky, you'll see a big pile with a "Free" sign on them, but you can also find them poking out of dumpsters or just piled up in a parking lot. If in doubt, always ask if you can take them. Most will be dirty and have a cracked or missing slat, but if you grab an extra you can use it for parts.

Making a Pallet Planter

After brushing off the dirt and renailing any loose boards, use some 80-grit sandpaper to clean up the areas that will be visible. Also round over rough, splintery edges. Paint or finish the outside of the pallet, if desired, but don't finish the inside if you're planting edibles.

Upcycling is a key part of any self-sufficient household, and pallets are some of the handiest candidates for the treatment. For this vertical planter project, the pallet is left intact and converted for use in a smaller area such as a patio, deck, or balcony. All part of turning every corner of the yard into a productive part of the whole.

Making the Planter

The process used to turn this pallet into a planter is a simple one and can be done by anyone with even very basic DIY skills. Decide which end of the pallet will be up, then cover openings at the back to keep dirt from falling out through the openings.

Gaps on the sides will be covered with a strip of metal flashing, an extra slat or a piece of wood cut to size. The back is covered with rubber pond liner or a double layer of black 6-mil poly (but make sure none of it is exposed to the sun, or it will decay).

Finally, nail a doubled-over strip of aluminum screen mesh across the bottom of the pallet to keep the dirt in, and then fill with topsoil. Tamp it down with a long stick to make sure the pallet fills up. Dirt will fall out the front at first, but will settle in at an angle behind each opening.

You can simply lean the planter against a wall in a sunny area if you prefer. However, you can also mount the planter on a southern or eastern-facing side of a garage, house, or outbuilding. If you prefer to put the planter out in the garden or in a sunny spot in the yard, you can screw stakes to each side or front and back, and then secure it in the ground so that it doesn't fall over when bumped or on a windy day.

If space is tight, you can even wire it to a balcony or deck railing. Just be sure the railing is solid because when you water the plants and soil in the planter, it can become fairly heavy.

Pallet Planter Options

You can make individual planters using the same method described here, but by cutting the pallet into sections. You can also decorate the pallet to better suit the style of the yard. As long as you don't get any paint inside, you can paint the outside in different colors or even stain the wood if you prefer.

Planting Your Pallet Planter

As you examine your completed pallet planter, one question immediately comes to mind: won't all the dirt fall out the front? Well, if you did not plant any plants in it the answer is yes, it will. However, you are relying on the plant roots and the 2 × 2 shelves you installed to hold things together. To add plants, lay the planter flat on its back and plan to keep it that way for a couple of weeks. Pack the gaps full of potting soil (potting soil is pre-fertilized, unlike topsoil) and then pack in as many seedlings as you can fit. Water the plants for a couple of weeks so the roots can establish. Then, tip the planter up against the wall in position. A little soil may trickle out initially, but you should find that everything holds together nicely.

BEST PALLET PLANTS

Not every vegetable or fruit will be at home in a pallet planter, but it can serve as the ideal location for several garden favorites.

Strawberries. Just as in a strawberry pot, these plants are at home with the shallow soil in the planter "pockets." The plants are easy to work with and, once settled in the planter, will produce a crop for several years.

Leaf lettuces. The plants will spread out in the pockets and, if kept watered and reasonably cool (you may need to shade the planter during the hottest part of the day), they will thrive in the planter.

Herbs. A pallet planter is idea for an herb garden. Because you only need a small amount of each herb, the planter can support an entire kitchen herb garden. It can also be set against a wall right outside a kitchen door, making the herbs incredibly convenient for harvesting.

HOW TO BUILD A PALLET PLANTER

TOOLS & MATERIALS

Cedar 2 × 2s, cut to fit
Cordless drill and bits
Screwdriver or pry bar
2" deck screws
Staple gun and staples
Pond liner or 6-mil poly
Scrap wood
Aluminum screen mesh
Eye and ear protection
Work gloves

Cut the 2 × 2 shelves to the length for the cavities in your pallet and fit them in on the lower edge of each slat. Predrill and toenail at each end to hold them in place, then drive an additional screw into the cedar from the front. Pallet wood is hard, so drill pilot holes for all screws. **NOTE:** Most pallets have a good face and a bad face. Be sure that you are creating your planter so the face with the nicer decking will point away from the wall you are installing it against.

Stiffen the pallet and provide a surface for attaching the liner. Fill the gaps on the back side with wood scraps of roughly the same thickness as the back boards.

Measure and cut pond liner or poly sheeting for the back. Fold the liner onto the sides and staple it in place on the sides and back. You can use heavy black poly sheeting to make the liner, but for a more durable material use roll rubber. Staple aluminum screen mesh across the bottom of the pallet.

Garden Projects

12 | Raised Beds

Raised garden beds offer several advantages over planting at ground level. When segregated, soil can be amended in a more targeted way to support high density plantings. Also, in raised garden beds, soil doesn't suffer compaction from foot traffic or machinery, so plant roots are free to spread and breathe more easily. Vegetables planted at high densities in raised beds are placed far enough apart to avoid overcrowding, but close enough to shade and choke out weeds. In raised beds, you can also water plants easily with soaker hoses, which deliver water to soil and roots rather than spraying leaves and inviting disease. And if your plants develop a fungus or another disorder, it is easier to treat and less likely to migrate to other plants in a raised bed situation.

Raised garden beds can be built in a wide variety of shapes and sizes, and can easily be customized to fit the space you have available on your property. Just make sure you can reach the center easily. If you can only access your raised bed from one side, it's best to build it no wider than 3 feet. Beds that you can access from both sides can be as wide as 6 feet, as long as you can reach the center. You can build your raised bed as long as you'd like.

Raised garden beds can be built from a wide variety of materials: 2× lumber, 4 × 4 posts, salvaged timbers, even scrap metals and other recycled goods. Make sure any lumber you choose (either new or salvaged) hasn't been treated with creosote, pentachlorophenol, or chromated copper arsenic (CCA). Lumber treated with newer, non-arsenate chemicals at higher saturation levels is rated for ground contact and is also a safe choice for bed frames. Rot-resistant redwood and cedar are good choices that will stand the test of time. Other softwoods, including pine, tamarack, and cypress, will also work, but can be subject to rot and may need to be replaced after a few years.

BED POSITIONS

If you're planting low-growing crops, position the bed with a north-south orientation, so both sides of the bed will be exposed to direct sunlight. For taller crops, position the bed east-west.

Raised garden beds make great vegetable gardens—they're easy to weed, simple to water, and the soil quality is easier to control, ensuring that your vegetable plants yield bountiful fresh produce. Your garden beds can be built at any height up to waist-level. It's best not to build them much taller than that, however, to make sure you can reach the center of your bed.

Garden Projects

VEGETABLE PLANT COMPATIBILITY CHART

VEGETABLE	LOVES	INCOMPATIBLE WITH	PLANTING SEASON
Asparagus	Tomatoes, parsley, basil		Early spring
Beans (bush)	Beets, carrots, cucumbers, potatoes	Fennel, garlic, onions	Spring
Cabbage & broccoli	Beets, celery, corn, dill, onions, oregano, sage	Fennel, pole beans, strawberries, tomatoes	Spring
Cantaloupe	Corn	Potatoes	Early summer
Carrots	Chives, leaf lettuce, onion, parsley, peas, rosemary, sage, tomatoes	Dill	Early spring
Celery	Beans, cabbages, cauliflower, leeks, tomatoes		Early summer
Corn	Beans, cucumbers, peas, potatoes, pumpkins, squash		Spring
Cucumbers	Beans, cabbages, corn, peas, radishes	Aromatic herbs, potatoes	Early summer
Eggplant	Beans	Potatoes	Spring
Lettuce	Carrots, cucumbers, onions, radishes, strawberries		Early spring
Onions & garlic	Beets, broccoli, cabbages, eggplant, lettuce, strawberries, tomatoes	Peas, beans	Early spring
Peas	Beans, carrots, corn, cucumbers, radishes, turnips	Chives, garlic, onions	Early spring
Potatoes	Beans, cabbage, corn, eggplant, peas	Cucumber, tomatoes, raspberries	Early spring
Pumpkins	Corn	Potatoes	Early summer
Radishes	Beans, beets, carrots, cucumbers, lettuce, peas, spinach, tomatoes		Early spring
Squash	Radishes	Potatoes	Early summer
Tomatoes	Asparagus, basil, carrots, chive, garlic, onions, parsley	Cabbages, fennel, potatoes	Dependent on the variety
Turnips	Beans, peas		Early spring

Companion Planting

The old adage is true—some vegetables do actually get along "like peas and carrots." Some species of vegetables are natural partners that benefit from each other when planted close. On the other hand, some combinations are troublesome, and one plant will inhibit the growth of another. You can plant these antagonists in the same garden—even in the same raised garden bed—just don't place them side by side. Use the table on the previous page to help you plan out your raised garden beds to ensure that your plants grow healthy, strong, and bear plentiful fruit.

WATERING RAISED BEDS

When the soil inside the planting bed pulls away from the edges of the bed, it's time to water. The best time of day to water is in the late afternoon or early to mid-morning. Avoid watering in midday, when the sun is hottest and water will quickly evaporate, or near sundown or at night, when too much moisture in the soil can cause mold and fungus to grow.

Start with Healthy Soil

The success or failure of any gardening effort generally lies beneath the surface. Soil is the support system for all plants—it provides a balanced meal of the nutrients that plants' roots need to grow deep and strong. If you plan to fill your raised beds with soil from your property, it's a good idea to have the soil tested first to assess its quality. Take a sample of your soil and submit it to a local agricultural extension service—a basic lab test will cost you between $15 and $25 and will give you detailed information about the nutrients available in your property's soil. Mixing soil from your property with compost, potting soil, or other additives is a smart and inexpensive way to improve its quality. After you've filled your beds with soil, add a 3-inch layer of mulch to the top to lock in moisture and keep your good soil from blowing away in strong winds. Lawn clippings, wood, bark chips, hay and straw, leaves, compost, and shredded newspaper all work well as mulch materials.

Sprinklers with high, arching spray patterns are afflicted by excess water evaporation, but if you choose a small, controllable sprinkler with a water pattern that stays low to the ground you can deliver water to your raised bed with minimal loss.

HOW TO BUILD A RAISED BED

This basic but very sturdy raised bed is made with 4 × 4 landscape timbers stacked with their ends staggered in classic log-cabin style. The corners are pinned together with 6" galvanized spikes (or, you can use timber screws). It is lined with landscape fabric and includes several weep holes in the bottom course for drainage. Consider adding a 2 × 8 ledge on the top row. Corner finials improve the appearance and provide hose guides to protect the plants in the bed.

Create an outline around your garden bed by tying mason's string to the stakes. Use a shovel to remove the grass inside the outline, then dig a 4"-wide trench for the first row of timbers around the perimeter. Lay the bottom course of timbers in the trenches. Where possible, add or remove soil as needed to bring the timbers to level—a level bed frame always looks better than a sloped one. If you do have a significant slope to address, terrace the beds.

Add the second layer, staggering the joints. Drill pilot holes at the corners and drive 6" galvanized spikes (or 6 to 8" timber screws) through the holes—use at least two per joint. Continue to build up layers in this fashion, until your bed reaches the desired height.

Line the bed with landscape fabric to contain the soil and help keep weeds out of the bed. Tack the fabric to the lower part of the top course with roofing nails. Some gardeners recommend drilling 1"-dia. weep holes in the bottom timber course at 2' intervals. Fill with a blend of soil, peat moss, and fertilizer (if desired) to within 2 or 3" of the top.

HOW TO BUILD A RAISED BED FROM A KIT

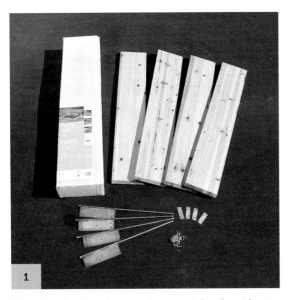

Raised garden bed kits come in many styles. Some have modular plastic or composite panels that fit together with grooves or with hardware. Others feature wood panels and metal corner hardware. Most kits can be stacked to increase bed height.

On a flat surface, assemble the panels and corner brackets (or hinge brackets) using the included hardware. Follow the kit instructions, making sure all corners are square.

Set the box down, experimenting with exact positioning until you find just the spot and angle you like. Be sure to observe the sun over an entire day when choosing the sunniest spot you can for growing vegetables. Cut around the edges of the planting bed box with a square-nose spade, move the box, and then slice off the sod in the bed area.

Set the bed box onto the installation site and check it for level. Add or remove soil as needed until it is level. Stake the box to the ground with the provided hardware. Add additional box kits on top of or next to the first box. Follow the manufacturer's instructions for connecting the modular units. Line the bed or beds with landscape fabric and fill with soil to within 2" or so of the top.

Garden Projects

BUILDING A RAISED BED WITH REMOVABLE TRELLIS

CUTTING LIST

KEY	NO.	PART	DIMENSION	MATERIAL
A	2	Side	1½ × 5½ × 72"	2 × 6 cedar
B	2	End	1½ × 5½ × 36"	2 × 6 cedar
C	2	Upper vertical	1½ × 60"	PVC pipe
D	2	Lower vertical	1½ × 12"	PVC pipe
E	2	Crosspiece	1½ × 34"	PVC pipe

TOOLS & MATERIALS

(2) 1½" × 10' PVC pipe
(2) 1½" PVC 90elbows
(2) 1½" PVC T-fittings
Heavy jute or hemp twine
Pipe straps for 1½" PVC (4 screw type)
Metal inside corners
Deck screws 1¼", 2½"
Tape measure
Cordless drill and bits
Hacksaw or miter saw
Sandpaper
Scissors or utility knife
Eye and ear protection
Work gloves

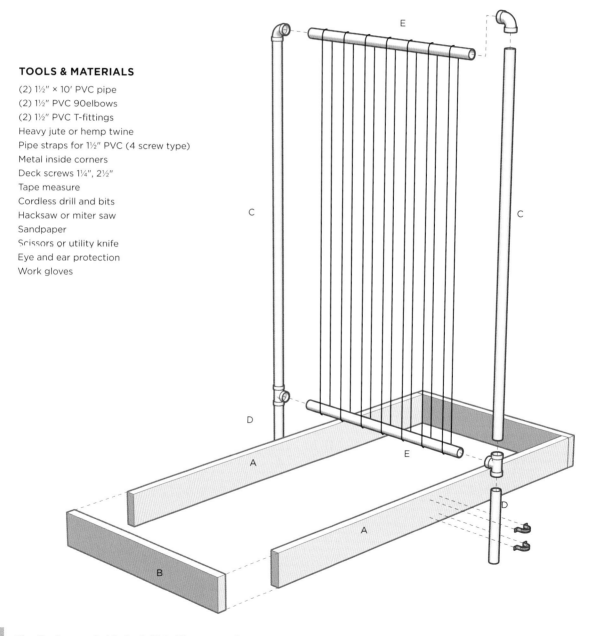

The Beginners Guide to Self Sufficency Projects for the Home

HOW TO BUILD A RAISED BED WITH A REMOVABLE TRELLIS

Start by assembling the raised-bed box, reinforcing the joints with metal inside corners. Add a center divider to keep the sides from spreading apart if you decide to make this project longer than 6'. Even if it is shorter, the divider is still a good precaution to help prevent warping.

Cut 12"-long pieces of 1½" PVC tubing. Attach them to the outsides of the planter box, near the middle. Use emery paper or sandpaper to remove the burrs and smooth the cut ends of pipe. Draw a perpendicular line where the pipe will go, using a square. Strap the pieces to the outsides with two pipe straps each. Fasten one strap with two screws, but leave the other strap loose until you put the upper vertical PVC on and can check it for plumb.

Add a T-fitting to the top end of each pipe. Measure between the hubs of the T-fittings to the insides of the sockets. Cut a piece of 1½" PVC pipe to this length and sand the cut edges smooth; this is the bottom crosspiece. Remove both Ts, fit the piece into the middle hubs of the Ts so the ends of the pipe bottom out in the fittings. Then replace the Ts.

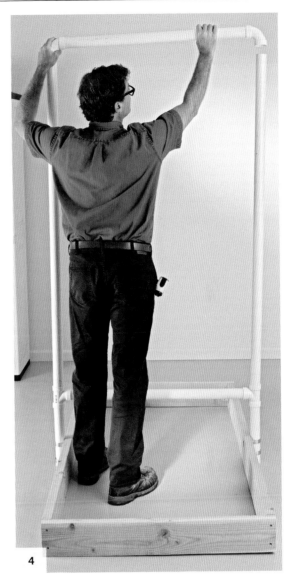

Add the uprights and attach the top crosspiece with elbows. Ensure the pipes are plumb, then secure the bottom straps. Move the planter into your yard or garden, line it with a thick layer of old newspaper or landscape fabric, and fill it with planting medium. Tie jute or hemp twine between the crosspieces so that climbing plants have something to grab onto. When winter comes, you can disassemble the PVC and store it away until spring.

Garden Projects

13 | Cold Frame

An inexpensive foray into greenhouse gardening, a cold frame is practical for starting plants six to eight weeks earlier in the growing season and for hardening off seedlings. Basically, a cold frame is a box set on the ground and topped with glass or plastic. Although mechanized models with thermostatically controlled atmospheres and sash that automatically open and close are available, you can easily build a basic cold frame yourself from materials you probably already have around the house.

The back of the frame should be about twice as tall as the front so the lid slopes to a favorable angle for capturing sunrays. Build the frame tall enough to accommodate the maximum height of the plants before they are removed. The frame can be made of brick, block, plastic, wood, or just about any material you have on hand. It should be built to keep drafts out and soil in.

If the frame is permanently sited, position it facing south to receive maximum light during winter and spring and to offer protection from wind. Partially burying it takes advantage of the insulation from the earth, but it also can cause water to collect and the direct soil contact will shorten the lifespan of the wood frame parts. Locating your frame near a wall, rock, or building adds additional insulation and protection from the elements. **TIP:** The ideal temperature inside is 65 to 75 degrees Fahrenheit during the day and 55 to 65 degrees Fahrenheit at night. Keep an inexpensive thermometer in a shaded spot inside the frame for quick reference. A bright spring day can heat a cold frame to as warm as 100 degrees Fahrenheit, so prop up or remove the cover as necessary to prevent overheating. And remember, the more you vent, the more you should water. On cold nights, especially when frost is predicted, cover the box with burlap, old quilts, or leaves to keep it warm inside.

Starting plants early in a cold frame is a great way to get a head start on the growing season. A cold frame is also a great place for hardening off delicate seedlings to prepare them for transplanting.

Garden Projects

A cold frame should only be used during the early, cooler days of the growing season when delicate seedlings need that extra protection, and for late-season frost protection. Once the warmer weather arrives and the plants are established, remove and relocate the cold frame. Ongoing usage will overheat and kill the plants. And while the clear acrylic lid on this cold frame is desirable because it is safer to work with and use than glass, too much heat buildup can cause the acrylic to warp.

BUILDING A COLD FRAME

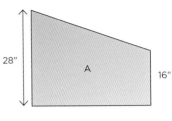

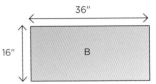

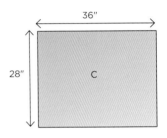

PLYWOOD CUTTING DIAGRAM

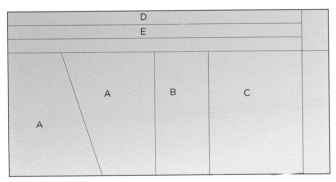

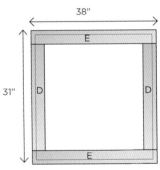

CUTTING LIST

KEY	PART	NO.	SIZE	MATERIAL
A	Side	2	¾ × 28 × 36"	Ext. Plywood
B	Front	1	¾ × 16 × 36"	Ext. Plywood
C	Back	1	¾ × 28 × 36"	Ext. Plywood
D	Lid frame	2	¾ × 4 × 31"	Ext. Plywood
E	Lid frame	2	¾ × 4 × 38"	Ext. Plywood
F	Cover	1	⅛ × 37 × 38"	Plexiglas

TOOLS & MATERIALS

- (2) 3 × 3" butt hinges (ext.)
- (2) 4" utility handles
- (4) Corner L-brackets (¾ × 2½")
- (1) ¾" × 4 × 8' Plywood (Ext.)
- ⅛ × 37 × 38" clear Plexiglas
- Exterior caulk/adhesive
- Exterior wood glue
- Eye and ear protection
- Exterior paint in darker color
- 2" deck screws
- #8 × ¾" wood screws
- Circular saw
- Drill/driver
- Pipe or bar clamps
- Straightedge cutting guide
- Work gloves

Garden Projects 91

HOW TO BUILD A PLYWOOD COLD FRAME

1 Cut the parts. This project, as dimensioned, is designed to be made entirely from a single 4 × 8 sheet of plywood. Start by cutting the plywood lengthwise to make a 36"-wide piece. **TIP:** Remove material in 4" wide strips and use the strips to make the lid frame parts and any other trim you may want to add.

2 Trim the parts to size with a circular saw or jigsaw and cutting guide. Mark the cutting lines first (see diagram, page 91).

3 Assemble the front, back, and side panels into a square box. Glue the joints and clamp them together with pipe or bar clamps. Adjust until the corners are square.

4 Reinforce the joints with 2" deck screws driven through countersunk pilot holes. Drive a screw every 4 to 6" along each joint.

Make the lid frame. Cut the 4"-wide strips of ¾" plywood reserved from step 1 into frame parts. Assemble the frame parts into a square 38 × 39" frame. There are many ways to join the parts so they create a flat frame. Because the Plexiglas cover will give the lid some rigidity, simply gluing the joints and reinforcing with an L-bracket at each inside corner is adequate structurally.

Paint the box and the frame with exterior paint, preferably in an enamel finish. A darker color will hold more solar heat.

Lay thick beads of clear exterior adhesive/caulk onto the tops of the frames and then seat the Plexiglas cover into the adhesive. Clean up squeeze-out right away. Once the adhesive has set, attach the lid with butt hinges and attach the handles to the sides.

Move the cold frame to the site. Clear and level the ground where it will set. Some gardeners like to excavate the site slightly.

Garden Projects

14 | Planting Trees

Wind saps heat from homes, forces snow into burdensome drifts, and can damage more tender plants in a landscape. To protect your outdoor living space, build an aesthetically pleasing wall—a "green" wall of tress and shrubs—that will cut the wind and keep those energy bills down. Windbreaks are commonly used in rural areas where sweeping acres of land are a runway for wind gusts. But even those on small, suburban lots will benefit from strategically placing plants to block the wind.

Essentially, windbreaks are plantings or screens that slow, direct, and block wind from protected areas. Natural windbreaks are comprised of shrubs, conifers, and deciduous trees. The keys to a successful windbreak are: height, width, density, and orientation. Height and width come with age. Density depends on the number of rows, type of foliage, and gaps. Ideally, a windbreak should be 60 to 80 percent dense. (No windbreak is 100 percent dense.) Orientation involves placing rows of plants at right angles to the wind. A rule of thumb is to plant a windbreak that is 10 times longer than its greatest height. And keep in mind that wind changes direction, so you may need a multiple-leg windbreak.

WINDBREAKS

Trees or shrubs planted in a row are aesthetically pleasing, plus they create a windbreak that muffles noise, provides privacy, and deflects harsh winds and drifting snow from your home and yard. Plant the windbreak at a right angle to the prevailing winds.

A stand of fast-growing trees, like these aspens, will create an effective windbreak for your property just a few years after saplings are planted.

Garden Projects

WINDBREAK BENEFITS

Windbreaks deliver many benefits to your property.

Energy conservation: reduce energy costs by 20 to 40 percent.

Snow control: single rows of shrubs function as snow fences.

Privacy: block a roadside view and protect animals from exposure to passers-by.

Noise control: muffle the sound of traffic if your pasture or home is near a road.

Aesthetic appeal: improve your landscape and increase the value of your property.

Erosion control: prevent dust from blowing; roots work against erosion.

Evergreens and deciduous trees both are effective windbreaks. These balsam trees grow only about 1' per year, but some faster-growing species exceed 10' per year in new growth.

TOOLS & MATERIALS
Shovel
Garden hose
Utility knife
Trees
Soil amendments (as needed)
Eye and ear protection
Work gloves

HOW TO BUILD A PLANT A WINDBREAK

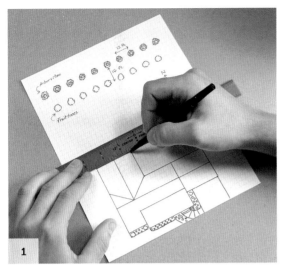

Before you pick up a shovel, draw a plan of your windbreak, taking into consideration the direction of the wind and location of nearby structures. Windbreaks can be straight lines of trees or curved formations. They may be several rows thick, or just a single row. If you only have room for one row, choose lush evergreens for the best density. Make a plan.

Once you decide on the best alignment of trees and shrubs, stake out reference lines for the rows. For a three-row windbreak, the inside row should be at least 75' from buildings or structures, with the outside row 100 to 150' away. Within this 25 to 75' area, plant rows 16 to 20' apart for shrubs and conifers and no closer than 14' for deciduous trees. Within rows, space trees so their foliage can mature and eventually improve the density.

Dig holes for tree root balls to the recommended depth (see pages 53–55). Your plan should arrange short trees or shrubs upwind and taller trees downwind. If your windbreak borders your home, choose attractive plants for the inside row and buffer them with evergreens or dense shrubs in the second row. If you only have room for two rows of plants, be sure to stagger the specimens so there are no gaps.

Plant the trees in the formation created in your plan. Here, a row of dwarf fruit trees is being planted in front of a row of denser, taller evergreens (*Techny Arborvitae*).

Garden Projects

SECTION 4

Food Prepartion and Preservation

We depend on our food, just as we do our air and water, to be healthy and wholesome. But increasingly, we have no idea where our food comes from and what is done to it before it reaches us. The farther we move from the source of our food, the more it needs to be processed to get to us. Although shelf life is increased, much can be lost.

That's why there's something wholly satisfying about raising and storing your own food. It's not just that it's amazingly healthy—that much should be obvious. More than that, it's that you're taking steps to ensure that there are no toxins in your family's food. You're also taking empty calories out of your diet—like when you replace soda with fresh-pressed cider. But beyond health, there's the benefit of more flavorful edibles.

Grow heirloom tomatoes and you've given yourself the gift of rich, beautiful, incredibly tasty vegetables that can't be rivaled by what's on offer at the local grocery. Can and preserve part of that harvest, and you'll be enjoying homemade marinara and stews through the winter, while other people are buying factory-canned soups full of preservatives and who knows what else.

But there's another side to self-sufficient food growing, harvesting, and preparation. Once you get outside and start pressing cider, drying fruit, or making your own cheese, you'll discover the vastly enjoyable pleasure of crafting your own wholesome food. The act of walking down a grocery store aisle simply can't compete with slicing up your own fresh-picked cabbage to make a family-favorite cole slaw, or hanging herbs just cut from your garden to make dried additions to your next soup. Get in the habit of tending your plants and working with the harvest they produce, and it quickly becomes the farthest thing from work.

15 Preserving Your Bounty

Preserving garden produce does not need to be an overwhelming task, but it is important to think through your strategy before you get started. To understand how to preserve food, it's important first to understand why fresh fruits and vegetables spoil and decay. There are two main culprits: First, external agents, such as bacteria and mold, break down and consume fresh food. Second, naturally occurring enzymes—the very same ones that cause fruits and veggies to ripen—are also responsible for their decay. Canning, freezing, drying, and cold storage (along with salt-curing/smoking and making fruit preserves) are all ways to slow down or halt these processes while retaining (to varying degrees) nutrition and taste.

Each method for preserving food has its strengths and weaknesses. It's important to weigh these carefully and decide which preservation method is the best fit for your garden's needs and your lifestyle.

PRESERVING METHODS

Canning. Canning changes the taste of foods and results in some vitamin loss, but is versatile and can be used to preserve many different kinds of foods for lengthy periods of time.

Cellaring. Live storage preserves produce with minimal effect on taste or nutritional value, but it only makes sense for a few foodstuffs. The fruits and vegetables that can be cellared for limited periods of time require a storage environment that must be carefully regulated.

Freezing and drying. Freezing and drying both retain a high percentage of vitamins and have a minimum effect on flavor, but only certain foods can be preserved with these methods and careful preparation and regulation is extremely important. Freezing can damage the cellular structure of fruits and vegetables, causing an unpleasant mushiness that makes them suitable only for dices and purees. Drying also changes the nature of fruits and vegetables; for example, drying fruits causes the caloric value to double or triple as the starches convert to sugars during the process.

Canning is one of the most popular forms of food preservation. The process is reliable and can be applied to a wide variety of vegetables.

CHOOSING THE BEST PRESERVATION METHOD

PRODUCE	CANNING	FREEZING	DEHYDRATION	LIVE STORAGE
Apples	✔		✔	✔
Asparagus	✔	✔		
Beans (green)	✔	✔		
Beans (lima)	✔	✔	✔	
Beets	✔			✔
Broccoli		✔	✔	
Brussel Sprouts			✔	✔
Cabbage				✔
Carrots	✔	✔		✔
Cauliflower		✔		
Celery				✔
Cherries	✔		✔	
Corn	✔	✔		
Cucumbers	✔ (with pickling)			
Onions				✔
Pears	✔		✔	✔
Peaches	✔	✔	✔	
Peas	✔	✔	✔	
Peppers (green)		✔	✔	
Peppers (hot)			✔	
Potatoes	✔			✔
Pumpkin	✔			✔
Radishes				✔
Spinach	✔	✔		
Squash (summer)	✔	✔	✔	
Squash (winter)	✔			✔
Strawberries		✔	✔	
Tomatoes	✔	✔	✔	

Preservation by Freezing

Freezing is the best way to preserve delicate vegetables. It's also a quick process that is perfectly suited to smaller batches of food. In this process, foods should be blanched to stabilize nutrients and texture, cooled to preserve color, packaged in an airtight container, and frozen as quickly as possible. Frozen food, if properly packaged and contained within a temperature-consistent frozen environment, can be preserved for as long as a year. Of course, the longer you wait to eat your food, the more it will break down, which results in a loss of taste and freshness. Food can also absorb ambient flavors in the freezer environment, negatively affecting the taste.

Generally, the colder you keep your freezer, the longer your frozen food will stay tasting fresh. For best results, use a chest freezer instead of the little box above your refrigerator. Although chest freezers are an investment, they maintain colder temperatures more consistently than your refrigerator's freezer. The ideal temperature for your chest freezer is −5 degrees Fahrenheit, and it should be no warmer than 0 degrees Fahrenheit. Even a few degrees above zero will cut the freezer life of your food in half.

Preservation by Drying

There are many advantages to dehydrating produce from your garden. Most dehydration methods require very little extra energy other than that already provided by the sun. Also, dehydrated foods, if prepared correctly, retain much of their original beauty and nutritional value. And since foods lose so much of their mass during the dehydration process, they do not require much space to store through the winter and can easily be rehydrated to taste delicious months after the harvest.

Dehydration is a food preservation technique that has been used for centuries all around the world. Removing 80 to 90 percent of the moisture in food, it halts the growth of spoilage bacteria and makes long-term storage possible. Warm, dry air moving over the exposed surface of the food pieces will absorb moisture from the food and carry it away. The higher the temperature of the air, the more moisture it will absorb, and the greater the air movement, the faster the moisture will be carried away.

Temperature matters a lot in food drying—air at a temperature of 82 degrees Fahrenheit will carry away twice as much moisture as air at 62 degrees Fahrenheit. This process also concentrates natural sugars in the foods. The faster the food is dried, the higher its vitamin content will be and the less its chance of contamination by mold. Extremely high temperatures, however, will cause the outside surface or skin of the food to shrivel too quickly, trapping moisture that may cause spoilage from the inside out. Exposure to sunlight also speeds up the drying process but can destroy some vitamins in foods.

Often, foods should be treated before drying. Blanching as you would for freezing is recommended for just about any vegetable (notable exceptions being onions and mushrooms). Some fruit and vegetables dry best if cut into pieces, whereas others should be left whole. Coating the produce can help preserve the bright color of skins. Many dipping mixtures may be used (consult a recipe book), but lemon juice is probably the most common.

Apples are a favorite fruit for drying because they retain so much of their flavor. Look for sweet varieties like Fuji. Core them and cut them into ⅛"-thick rings or slices for drying. Peeling is optional. Dip the apples in lemon juice immediately after cutting or peeling to prevent browning.

Preservation by Home Canning

Canning is a traditional method for preserving produce. It is not difficult to master, but it's important to pace yourself. Try not to plan more than one canning project a day to keep the work manageable and enjoyable. Also, make sure you are familiar with how to use your canning equipment safely, and that you have a reliable recipe to reference for each food you plan to can. Every fruit and vegetable has a different acidity and requires slightly different accommodations in the canning process.

To get started with canning, there are two main tools to become familiar with: a water bath canner and a pressure canner. Foods with high acidity, such as fruits (including tomatoes), can be canned in a boiling water bath. Less acidic foods, including most vegetables, and any combination of high- and low-acidity foods must be processed using a pressure canner. Water bath and pressure canners are NOT interchangeable, largely because they reach vastly different temperatures during their processes. Always make sure the canner you use is appropriate for the produce you're preserving and follow your canning recipe exactly.

Other tools you'll need include canning jars, measuring cups, a long-handled spoon, a funnel, a jar lifter, and cooking pots. Canning jars typically have two-piece metal lids: the metal band can be reused whereas the disc part of the lid cannot form an adequate seal more than once, and should be discarded after one use. Always inspect jars carefully before beginning. Check for nicks on the rim or cracks anywhere in the jar. Discard or repurpose any imperfect jars as they will not be able to form an adequate seal.

> **BEWARE OF BOTULISM**
>
> Home canning is perfectly safe if all instructions are followed exactly. However, if canning procedures are not followed, harmful bacteria can fester while your produce sits in your pantry waiting to be used. The best known of these bacteria is *clostridium botulinum*, which produces a potent toxin that is odorless, colorless, and fatal to humans in small amounts. Cases of botulism poisoning are rare, but to avoid this toxic substance, it's important to always follow the home canning recipe and procedures exactly. And if in doubt, throw it out.

Make sure you understand how to use your home canning equipment before you get started. Take time to read through the manual that comes with your canner, and make sure you use the right type of home canner for the fruit or vegetable you're planning to preserve.

The Home Canning Process

1. **Wash and heat the jars.** Immerse jars in simmering water for at least 10 minutes or steam them for 15 minutes. Heat jar lids (just the disc part) in a small saucepan of water for at least 10 minutes. Keep lids hot, removing one at a time as needed.

2. **Pack food in the jars.** Different packing methods are used for different types of produce. In cold packing, raw food is placed in a hot jar and then hot liquid is poured over the food to fill the jar. In hot packing, foods are precooked and poured into a hot jar immediately after removing them from the heat source.

3. **Watch your headspace.** Headspace is the amount of space between the rim of the jar

and the top of the food and is very important to making sure that your canning jars seal correctly. Always follow your recipe's directions—generally it's best to leave about 1 inch of headspace for low-acid foods, ½ inch for acidic foods, and ¼ inch for pickles, relishes, jellies, and juices.

4. **Remove air bubbles.** Insert a nonmetal spatula or chopstick and agitate the food to remove all air bubbles.

5. **Place the lid.** Clean the jar rim, then set a hot disc on the jar rim and screw on the band until you meet the initial point of resistance and no further.

6. **Heat.** Place jars on the rack in the water bath or pressure canner and process immediately. Follow the directions for your canner.

7. **Cool.** Allow the jars to cool slowly after processing—cooling too quickly can cause breakage. Typically, jars should cool along with the water they're submerged in, but follow the directions for your canner. Do not tighten the lids unless they are very loose. As the jars cool, you'll hear them "pop" when they are properly sealed. If the jar does not seal, refrigerate and eat within the next couple days.

8. **Clean and label.** After cooling and confirming the jar's seal, wash the outside of the jar and label with the content and date.

9. **Store.** Store in a cool, dark cupboard or pantry. If a jar loses its seal during storage (i.e., if the metal disc does not pop when you remove it), the food inside is not safe to eat. Dump it on the compost bin and try a different jar.

CANNED FOOD SAFETY QUIZ

Ask yourself the following 10 questions to determine if your home-canned food is safe to eat:

1. Is the food in the jar covered with liquid and fully packed?
2. Has proper headspace been maintained?
3. Is the food free from moving air bubbles?
4. Does the jar have a tight seal?
5. Is the jar free from seepage and oozing from under the lid?
6. Has the food maintained a uniform color?
7. Is the liquid clear (not cloudy) and free of sediment?
8. Did the jar open with a clear "pop" or "hiss" and without any liquid spurts?
9. After opening, was the food free of any unusual odors?
10. Is the food and underside of the lid free of any cottonlike growths?

If you can answer "yes" to all of the questions, your food is probably safe. That said—if you have even a small suspicion that a jar of food is spoiled—dump it in the compost bin. Never, under any circumstances, taste food from a jar you suspect may have spoiled or lost its seal. Botulism spores have no odor, cannot be seen by our eyes, and can be fatal, even in small doses.

When drying produce in the oven, leave the door slightly ajar to allow moisture to escape, and carefully monitor temperature to ensure the oven doesn't heat to over 150 degrees Fahrenheit.

Drying Produce Indoors

Drying vegetables indoors allows you to carefully control the drying conditions and offers more protection from insects and changes in weather. An electric food dehydrator appliance is the simplest choice for indoor drying. If you don't have a dehydrator, the next best option is in or around your oven, although any hot, dry area will do—possibly even your attic or the area around a heater or cookstove.

If you plan to dry produce in your oven, keep in mind that the process typically takes 8 to 12 hours. Preheat your oven and check that it can maintain a temperature of 130 to 145 degrees Fahrenheit for at least an hour—some ovens have a difficult time holding low temperatures like this, and going over 150 degrees Fahrenheit can be disastrous for drying produce. Wash and prepare the food, then spread food in single layers on baking sheets, making sure the pieces do not touch. Place the sheets directly on the oven racks, leaving at least 4 inches above and below for air circulation. Also, make sure to leave the oven door slightly ajar to allow moisture to escape. Rearrange the trays and shift food from time to time to ensure even drying.

You may also dry food on your oven's range by creating a chafing dish. To create a chafing dish on your range, you'll need two baking trays: The first must be large enough to cover all burners and hold a 3-inch-deep reservoir of water. The second tray should fit on top of the first. Fill the bottom tray with water and set all burners to low heat. Throughout the process, refill the reservoir periodically to make sure food doesn't burn, and move/turn food as necessary to ensure even drying. Place a fan nearby to keep the air moving around the room, which will help carry moisture away from the food more quickly.

How Long Does Dehydration Take?

Drying times vary considerably—from a few hours to many days, depending on the climate, humidity, drying method, and the moisture content of the food you're dehydrating. Generally, fruit is done drying if it appears leathery and tough and no

An electric produce dehydrator can dry large quantities of fruits or vegetables in a sanitary environment. The stackable trays allow you to match the appliance's drying capacity to your needs each time you use it.

moisture can be squeezed from it. Vegetables should be so brittle and crisp that they rattle on the tray. To check for completed dehydration, you can also check the food's weight before and after the process. If the food has lost half its weight, it is two-thirds dry, so you should continue to dry for half the time you've already dried.

To double-check that your food is dry, place it in a wide-rimmed, open-topped bowl covered with cheesecloth fastened with a rubber band. Place the bowl in a dry place, and keep the food in the bowl for about a week. Stir it a couple times a day—if any moisture or condensation appears, you should continue to dehydrate.

Pasteurization & Storage

Regardless of the drying method used, food should be pasteurized before storage to ensure that there are no insect eggs or spoilage microorganisms present. To pasteurize, preheat the oven to 175 degrees Fahrenheit. Spread dried food 1 inch deep on trays and bake in the oven for 10 to 15 minutes. Dried food is best stored in clean glass jars or plastic bags in a cool, dry place. Never store dried food in metal containers and carefully monitor the humidity of the storage environment. Containers should have tight-fitting lids and should be stored in a dark, dry place with an air temperature below 60 degrees Fahrenheit.

> ### ENJOYING YOUR DEHYDRATED FOOD
>
> Many foods are delicious and ready to eat in their dried forms—especially tomatoes and berries. But dried food can also be rehydrated before eating. To rehydrate food, pour boiling water over it in a ratio of 1 1/2 cups of water to 1 cup of dried food, then let the food soak until all the water has been absorbed. You may also steam fruit or vegetables until rehydrated. Rehydrated vegetables should be cooked before eating, whereas rehydrated fruits are acceptable to eat without cooking after rehydration.

After drying and before storing, use your oven to heat fruits and vegetables to a high enough temperature to kill bacteria and related contaminants. About 15 minutes at 175° F will suffice for most produce, provided it is not in layers over 1" deep. Oven-drying takes about a half day at 140° F or so.

16 | Solar Dryer

A solar dryer is a drying tool that makes it possible to air-dry produce even when conditions are less than ideal. This dryer is easy to make, lightweight, and is space efficient. The dryer makes a great addition to your self-sufficient home, allowing you to use your outdoor space for more than gardening. The dryer, which is made of cedar or pine, utilizes a salvaged window for a cover.

But you will have to adjust the dimensions given here for the size window that you find. The key to successful solar drying is to check the dryer frequently to make sure that it stays in the sun. If the air becomes cool and damp, the food will become a haven for bacteria. In a sunny area, your produce will dry in a couple days. Add a thermometer to the inside of your dryer box, and check on the temperature frequently—it should stay between 95 and 145 degrees Fahrenheit. You may choose to dry any number of different vegetables and fruits in the dryer, such as:

- Tomatoes
- Bananas
- Squash
- Apples
- Peppers

A SIMPLE DRYER FOR A HOT CLIMATE

If you live in an area with clean air and a sunny, hot, dry climate, you can simply load food onto a drying tray or rack and place it out in the sun on blocks so that air circulates around it. Cover the food with cheesecloth held a few inches off the food on sticks to keep insects away, and bring the tray indoor at nights. Drying will take two or three days.

An old glass window sash gets new life as the heat-trapping cover of this solar dryer.

BUILDING A SOLAR DRYER

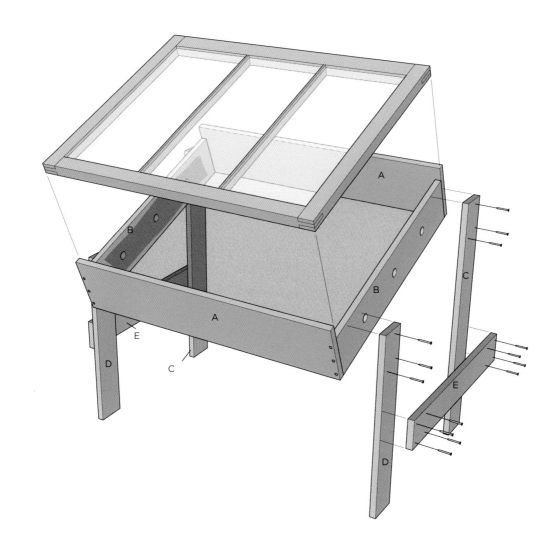

CUTTING LIST

KEY	PART	NO.	DIMENSION	MATERIAL
A	Front/back	2	¾ × 7½ × 34¾"	Cedar
B	Side	2	¾ × 5½ × 27⅞"	Cedar
C	Leg (tall)	2	¾ × 3½ × 30"	Cedar
D	Leg (short)	2	¾ × 3½ × 22"	Cedar
E	Brace	2	¾ × 3½ × 24"	Cedar

TOOLS & MATERIALS

1" spade bit
Circular saw
(1) 1 × 8" × 8'
(1) 1 × 6" × 8'
Eye protection
(2) 1 × 4" × 8'
Stapler
1¼" deck screws
Drill
Staples
Insect Mesh—fiberglass 28⅞ × 34¾"
Window sash
1½" galv. finish nails
Brad nails
Eye and ear protection
Work gloves

HOW TO INSTALL A SOLAR DRYER

Assemble the box. Attach the wider boards for the frame by driving screws through the faces of the 1 × 8" boards into the ends of the 1 × 6" boards. There will be a difference in height between these pairs of boards so that the window sash can sit flush in the recess created.

Install the mesh. Staple the screen to the frame. Then tack the retainer strips over the screen to the frame with 3-4 brad nails per side. Trim off the excess mesh.

Build the stand. Attach each 24" board to a 30" board (in the back) and a 22" board (in the front) with 1¼" deck screws. Then attach the finished posts to the frame with three 1¼" deck screws in each post.

Drill three 1" holes for ventilation in each 1 × 6" board equally spaced along the length of the board, leaving 5" of room on each end for the posts. Staple leftover insect mesh behind the ventilation holes on the inside of the frame.

Finish the project by sliding the window sash into place.

17 | Root Vegetable Rack

Before the widespread use of refrigeration, one of the methods home gardeners used for preserving root vegetables was to spread them out on drying racks in cool, dry basements and root cellars. That basic method still works just fine, and it's a better than just dumping your bumper crop into bags or boxes, where they're more susceptible to mold and rot.

This classic drying rack can hold hundreds of potatoes, carrots, beets, and other garden vegetables and fruits, and it's dirt simple to build. Made from inexpensive common pine, you can leave it unfinished or wipe on a food-safe butcher block oil (make sure it says "Food-safe" or "Food-grade" on the label). You can easily make the rack larger or smaller than our version by changing a few lengths, or even make a smaller countertop version for your kitchen; just be sure to leave the gaps between the slats for air movement and to follow the design for the drawers. The drawers can be opened from either direction for easy access to the vegetables.

Because the drawers are made from individual slats instead of plywood, there are lots of pieces to cut, but most of them are the same length, and if you take a moment to set up a stop block and cut several pieces at a time, you can cut them all in minutes. A compressor and nail gun will speed up construction considerably, but you can also just use screws or hand-nail everything, as carpenters would have done back in the nineteenth century.

Keep your vegetables and apples dry and fresh with this classic pine drying rack.

BUILDING A ROOT VEGETABLE RACK

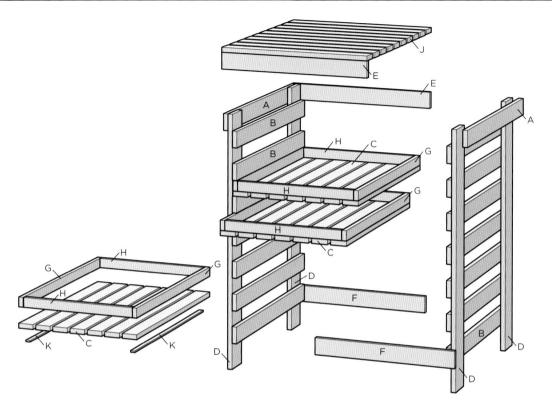

CUTTING LIST

KEY	NO.	PART	DIMENSION	MATERIAL
A	2	Top side trim	¾ × 2½ × 30"	Pine
B	14	Drawer support	¾ × 2½ × 30"	Pine
C	42	Drawer slat	¾ × 2½ × 30"	Pine
D	4	Leg	¾ × 3½ × 38"	Pine
E	2	Top front and back	¾ × 2½ × 23¾"	Pine
F	2	Bottom stretcher	¾ × 2½ × 22¼"	Pine
G	12	Drawer side	¾ × 1½ × 30"	Pine
H	12	Drawer front and back	¾ × 1½ × 19"	Pine
J	10	Top slat	¾ × 2½ × 23¾"	Pine
K	12	Drawer guide	¼ × ¾ × 28"	Pine screen mold

TOOLS & MATERIALS

Miter saw
Drill
Compressor and finish nailer or narrow crown stapler
Framing square
Clamps
Countersink bit

Wood glue
Self-tapping or drywall screws—1¼" and 2"
1½" finish nails (or narrow crown staples)
¾" brad nails
(21) 1 × 3 × 8' pine

(2) 1 × 4 × 8' pine
(7) 1 × 2 × 8' pine
(4) ¼ × ¾ × 8' pine screen mold
Sandpaper–150 grit
Eye and ear protection
Work gloves

HOW TO BUILD A ROOT VEGETABLE RACK

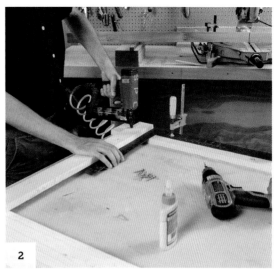

Cut 56 lengths of 1 × 3 and 12 lengths of 1 × 2 to 30" (Parts A, B, C, G). To ensure that all the pieces are the same, set up a stop block wide enough so that you can cut several pieces at once. One simple way to make a stop block is to screw a long, temporary 1 × 3 fence to the miter saw fence, mark 30" from the blade, then screw on a stop block. Move the stop block as needed to cut the rest of the pieces.

Cut the legs and glue and nail on the bottom 1 × 3 drawer support. Keep the 1 × 3 square to the legs as you work by clamping a square in place. Glue each joint, nail it in place (if you have a pneumatic nailer), and then reinforce each joint with two 1¼" screws. Predrill with a countersink bit to avoid splitting the wood if you are hand-nailing.

Using a straight piece of 1 × 3 as a spacer, add the rest of the 1 × 3 drawer supports. Check the distance to the top frequently to make sure you're still square. Glue and nail each joint. Repeat these steps to build the second side.

(continued)

HOW TO BUILD A ROOT VEGETABLE RACK (continued)

Stand the two sides up, top ends down, and join them with the top front and back pieces (E). Place the bottom stretchers (F) ¼" below the level of the first pair of drawer supports so that the drawer guides (K) on the bottom of the drawers will clear. Fasten all these joints with glue and predrilled screws for extra strength. Stand the drying rack right side up and set it aside.

Nail the 1 × 2 drawer sides together using a dab of glue to strengthen the joints. Make sure that the pieces are square.

Glue and nail the 1 × 3 bottom slats. Nail on the edge pieces first to keep the drawer square, then use ½" spacers (full ½", not ½" plywood, which is slightly less than ½") to space the rest of the slats.

Sand the drawer bottoms with 150-grit sandpaper so they slide smoothly in the runners. Also sand the edges and corners to give the piece a more finished look.

Nail drawer guides to the bottom of each drawer so that they will stay on the tracks. Space the guides ¾" from each edge, using a piece of 1 × 3 as a guide. Remember to use ¾" brads to nail the guides. The guides are 28" so that they can be set back from the edges an inch, which makes them almost invisible.

The top is optional, but it gives the drying rack a more finished look, plus it creates more storage space. Cut the pieces using the stop block, then glue and nail them in place. Space the slats about 11/16" apart, but you may need to adjust the spacing for the last few pieces.

SECTION 5

Homestead Amenities

The most important word in homestead is "home." After all, everything you're doing in a move toward self-sufficiency is aimed at making your home a better place to live. You can actually consider homesteading the improvement of your many homes—the world, your region, and the plot of land that holds that structure where you lay your head at night. This section is, in the final analysis, all about that last one.

As anyone knows, a home can either be warmly inviting, comforting and wonderful, or cold and off-putting, and never quite settled. The amenities in this chapter focus on moving your home toward the former rather than the latter. You see, when you make your home—including the inside and outside—more functional, you make it a nicer place to live.

You can consider these practical amenities. Having a handy place to stack and season your firewood, a countertop wine rack, or a solar heat panel is all about convenience. Projects like these also often provide viable alternatives to using expensive appliances that are energy hogs. Do a load of laundry in a manual laundry machine and you save electricity even while you discover the meditative bliss of losing yourself in 20 minutes of non-taxing labor performing an essential household task. Beyond convenience and the other rewards of these projects lies something else. Something that isn't quite as tangible, but to anybody who values self-sufficiency and a return to a simple life, is just as important. That is handmade craftsmanship.

Surrounding ourselves with conveniences isn't hard. A few hours spent at an appliance store can do that. But surrounding ourselves with conveniences that give us a connection to our home and remind us that yes, we can do for ourselves . . . well, those are simply the best type of conveniences and true homestead amenities.

18 | Solar Oven

There are many effective ways to make a solar cooker—one website devoted to the subject features dozens of photos of different types sent in by people from all around the world. The basic principle is so fundamental that it is easily adapted to a range of styles. We settled on this particular design mostly because it's low-cost to build, working with wood is often easier than manipulating metal, and the unit can cook about any meal you might need to make. However, it's easy enough to modify this design to suit your own food preparation needs.

The cooker is big enough to hold two medium-size pots. All the pieces are cut from one 8-foot 2 × 12 and a sheet of ¾-inch plywood. (The cooker would work just as well with ¼-inch plywood, but we used ¾-inch because it made it simpler to screw the corners and edges together.) The base is made from 1½-inch thick lumber for ease of construction and for the insulation value of the thicker wood, but thinner material would also work.

The foil we used was a type recommended for durability and resistance to UV degradation by an independent research institute. Unfortunately, it was expensive, and if you're just starting out, you may want to do a trial run with heavy-duty aluminum foil. Although foil looks a little dull, it actually reflects solar rays almost as well as specially polished mirrors.

In operation, the cooker is the height of simplicity. The sun's rays reflect off the foil sides and are concentrated at the base of the cooker, where they are absorbed by the black pot. The glass cover (or clear oven cooking bag) helps hold heat and moisture in the pot. The cooker should face the sun. Raise or lower the box depending on the time of year so that you catch the sunlight straight on. Shim the wire rack as needed to keep the pot level.

A solar cooker is an incredibly useful appliance that exploits the limitless energy in sunshine to cook meals large and small.

TYPES OF SOLAR COOKERS

There are many different types of solar cookers. Really, the only requirement is that the sun's rays be captured and focused on whatever is being cooked. Beyond that, the actual construction of the unit can be left to the imagination. However, the most common and popular types of solar cookers can be divided between three groups.

PARABOLIC CONCENTRATORS

Sometimes called curved concentrators, these are a more sophisticated design, but they accommodate only one cooking pot at a time in most cases. However, the shape effectively focuses the sun's rays much better than other types of solar cookers, resulting in higher cooking temperatures. These cook faster, but must be monitored more closely to ensure proper cooking. Crude versions can sometimes be adapted from retired satellite dishes or other parabolic devices.

BOX COOKERS

These are probably the most popular type of solar cooker because they are so easy to build. The shell can be even be made from found wood or other scavenged materials of odd sizes (which probably accounts for the prevalence of this type of cooker in the third world and impoverished areas). Basically, all you need is a box with a reflective surface inside, and a reflective lid. The box is positioned facing south, and opened at an angle that best directs the sunlight down into the cavity of the box. This type of cooker can be built small or large, is highly portable, and can be constructed to accommodate specifically the dimensions of the cookware that the user already owns.

PANEL COOKERS

These can be considered a hybrid of the other two styles of cookers. These are a fundamental design that is easy to construct and works well—sometimes better than a box cooker. They are available as kits from self-sufficiency and survivalist manufacturers, but with a little bit of thought and effort, one can easily be constructed from scratch. Be prepared to experiment with the angling of the panels to find exactly the orientation that will work best for your location and situation.

BUILDING A SOLAR OVEN

CUTTING LIST

KEY	NO.	PART	DIMENSION	MATERIAL
A	2	Base side	1½ × 11¼ × 19″	2 × 2 pine
B	2	Base end	1½ × 11¼ × 16″	2 × 2 pine
C	1	Bottom	¾ × 19 × 19″	¾″ ext. grade
D	1	Adjustable leg	¾ × 10 × 17″	¾″ ext. grade
E	1	Hood back	¾ × 20 × 33¾″	¾″ ext. grade
F	1	Hood front	¾ × 10 × 25¼″	¾″ ext. grade
G	2	Hood side	¾ × 20 × 31¼″	¾″ ext. grade
H	1	Lens	¼ × 17¼ × 17¼″	Tempered glass

TOOLS & MATERIALS

- Straightedge
- Circular saw
- Jigsaw or plunge router
- Tape measure
- Drill driver with bits
- Speed square
- Stapler
- #8 countersink bit
- ¾″ × 4 × 8′ BC or better plywood
- 2 × 12 × 8′ SPF SolaReflex foil or heavy-duty aluminum foil
- Bar clamps
- 1⅝″, 2½″ deck screws
- Clear silicone caulk
- Mid-size black metal pot with glass top
- Wire rack
- No-bore glass lid pulls
- ¼ × 2″ hanger bolts with large fender washers and wingnuts
- Sander
- Glue
- Eye and ear protection
- Work gloves

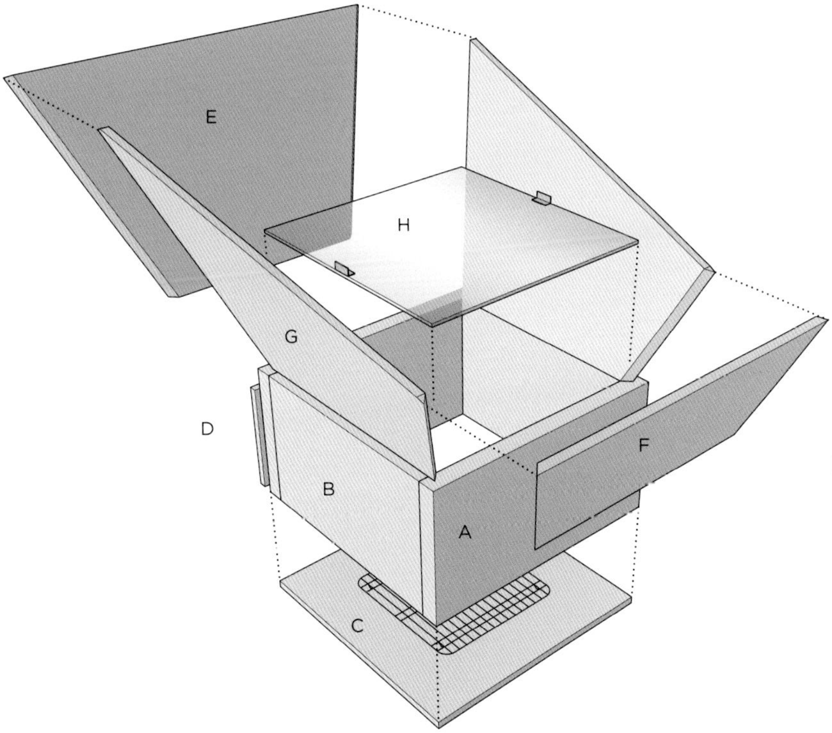

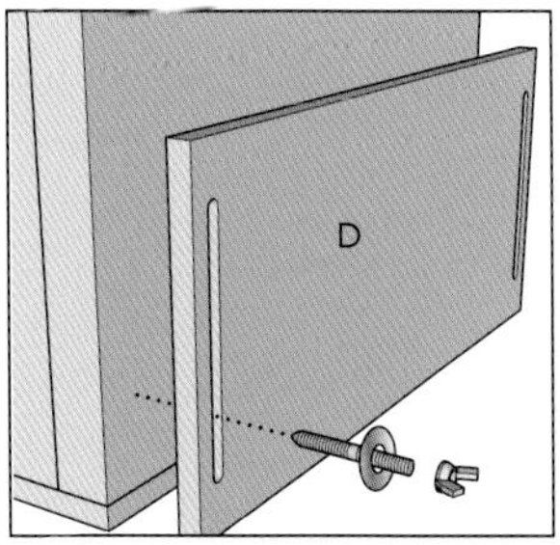

HOW TO BUILD A SOLAR OVEN

Cut the four 2 × 12 base pieces to length according to the cutting list. Arrange the base parts on a flat work surface and clamp them together in the correct orientation. Check for square and adjust as needed. Drill pilot holes and fasten the pieces together with 2½" deck screws.

Lay a sheet of plywood on the work surface with the better side facing up. Mark and cut the bottom first. Rest the full sheet of plywood on a couple of old 2 × 4s.

To create the panels that form the reflector, you'll need to make beveled cuts on the bottom and sides so the panels fit together squarely. Mark two 20 × 76" long pieces, measuring from the two factory edges so the waste will be in the middle. Set your circular saw base to 22½, then cut along the line you drew at 20". Cut the other piece starting from the opposite end of the plywood. You should end up with two mirror-image pieces.

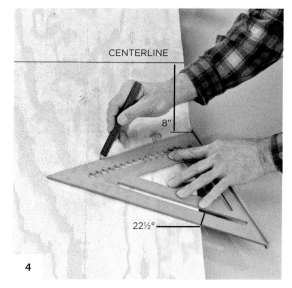

Reset your saw base to 0°, then cut each 20"-wide panel in half so you have four 20 × 38" panels, each with one beveled 38" edge. With the beveled edge facing up and closest to you, draw a centerline at 18" on each panel, then make marks on the beveled edges at 8" on both sides of the centerline. Position a speed square so it pivots at the 8" mark, then rotate the speed square away from the centerline until the 22½° mark on the speed square meets the top of the beveled edge. Draw a line and use a straightedge to extend the line to the other edge (the factory edge) of the plywood. Repeat at the other 8" mark, flipping the speed square and rotating it away from the centerline so the lines create a flat-topped triangle. Set the base of your circular saw at 40°, then cut along the angled lines. Mark and cut the remaining three panels in the same fashion.

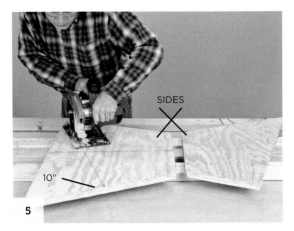

5 Finish cutting the reflector parts to final size and shape. **TIP:** After you've laid out your cutting lines, set the workpiece onto a pair of old 2 × 4s. Tack the workpieces to the 2 × 4s with finish nails, driven into the waste area of the panels. Keep the nails at least a couple of inches from any cutting line. Set your saw so the cutting depth is about ¼" more than the thickness of the workpiece and then make your cuts.

6 Assemble the reflector. Brace two of the reflector sides against a square piece of scrap plywood clamped to the work surface, then join the edges with screws driven into countersunk pilot holes. Repeat for the other two pieces. Join the two halves together with four screws at each corner, completing the reflector. The bottom edges should be aligned. The top edges won't match perfectly, so sand them smooth.

COMPOUND MITER CORNER CUTS

The sides of this solar cooker box are cut with the same basic technique used to cut crown molding. Instead of angling the crown against the miter saw fence in the same position it will be against the ceiling—a simple 45° cut that is easy to visualize—you have to make the compound cuts with the wood lying flat, which makes it mind-bendingly difficult to visualize the cut angles. For the dimensions of this cooker, a 40° bevel cut along the 22½° line will form a square corner. If you change the 22½° angle, the saw cut will also change.

If you remember your geometry, you can work all this out on paper. But bevel guides on circular saws are not very precise, and 40° on one saw might be more like 39° on a different brand; test cuts are the best way to get the angle right. Make the first cuts a little long and then try them out.

The easiest way to avoid a miscut is to lay all the pieces out with the bases lined up and the good side of the plywood up. Mark the 22½° lines for the sides, then cut the 40° angles on one edge of all four pieces. Next, flip the piece around and cut the 40° angle on the other side. Remember, the 40° cut should angle outwards from the good side of the plywood, and the pieces should all be mirror images.

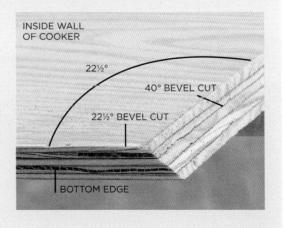

(continued)

HOW TO BUILD A SOLAR OVEN (continued)

7

Cut the adjustable leg with parallel slots so the leg can move up and down over a pair of hanger bolts, angling the cooker as necessary. Outline the slots so they are ⅜" wide (or slightly wider than your hanger bolt shafts). Locate a slot 2" from each edge of the adjustable leg. The slots should stop and start 2" from the top and bottom edges. Cut the slots with a jigsaw or a plunge router.

8

Screw the plywood bottom to the base. Set the adjustable leg against one side of the base, then drill guide holes and install the hanger bolts to align with the slots. Center the bolts at the same height: roughly 2½" up from the bottom of the box. Use large fender washers and wing nuts to lock the adjustable leg in position.

9

Fasten the reflector to the base with countersunk 2½" deck screws. Angle the drill bit slightly as you drill to avoid breaking the plywood edge. Use two screws per side.

10

Cut pieces of reflective sheeting to fit the sides of the reflector as well as the base. You can use heavy-duty aluminum foil, but for a sturdier option try solar foil. Cut the pieces large enough to overlap at the edges.

11

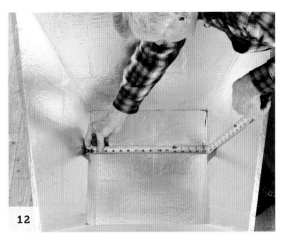

12

Take measurements to double-check the glass lid size. Ideally, the lid will rest about 1" above the top opening of the box. Order glass with polished edges. You can also use a clear plastic oven bag instead of the glass. Either will trap heat and speed up the cooking.

Glue the reflective sheeting inside the base and reflector, overlapping the sheets. Use contact cement or silicone caulk and staple the edges to reinforce the glue (use diluted white glue with a paintbrush instead of contact cement if you're using aluminum foil). Smooth out the reflective material as much as possible; the smoother the surface is, the better it will reflect light.

GETTING A HANDLE ON GLASS

Because it is virtually impossible to lift the glass lid from above, you'll need to install handles or pulls designed to attach to glass (available from woodworking hardware suppliers). The simplest of these require no drilling. You squeeze a bead of clear, 100 percent silicone into the U-channel of the lid handle, then slide the handle over the edge of the glass.

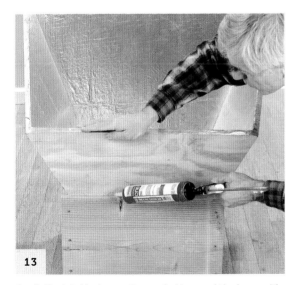

13

Caulk the joint between the angled top and the base with clear silicone caulk. Set a wire rack inside the oven to keep the cooking pot slightly elevated and allow airflow beneath it.

19 | Frame Loom

Weaving your own textiles can be incredibly relaxing, enjoyable, and fulfilling. It's a way to reclaim a heritage craft and create beautiful fabric pieces, from scarves to rugs. The trick is to build your skills on a small loom first, and then progress to a larger, more complicated loom. This naturally means starting small, but the idea is to build on your successes, and avoid the frustration that can come from handling a more sophisticated apparatus before you're ready.

Frame looms like the one in this project are ideal for the beginner or intermediate weaver. They are easy to handle, portable, and small enough that they don't take up much room. Just the same, you'll be able to create decorative fabric panels for hanging, small runner rugs, scarves, and even panels that can be sewn together to create more involved projects like a quilt.

The construction of this loom is simple and straightforward. The example here is a fairly standard size, but don't be afraid to resize the dimensions to suit your own needs and preferences. Just be careful not to make it too big or the frame will have a tendency to flex as you work, making the weaving more difficult. We've also included legs on this frame loom to make weaving more comfortable. Adjust the position so that you can sit comfortably and weave without excessive reaching or fumbling.

As the name implies, you begin by building a fundamental frame to which the supporting (or "warp") fibers will be secured. The "weft" threads that run horizontally are then woven through these warp threads. This is the basic process of any loom—the technique just becomes more involved the bigger and more complex the loom. The terms associated with loom weaving can be a little confusing to the beginner, so we've included a glossary to keep things straight. No matter what words you use, however, the result will be a fabric that you've created with your own hands—no fabric store or mill necessary!

Create marvelous handicrafts with a simple frame loom like this one. It's easy to use and will help develop your weaving skills, should you ever want to step up to a standalone loom and bigger textiles like rugs and bedspreads.

BUILDING A FRAME LOOM

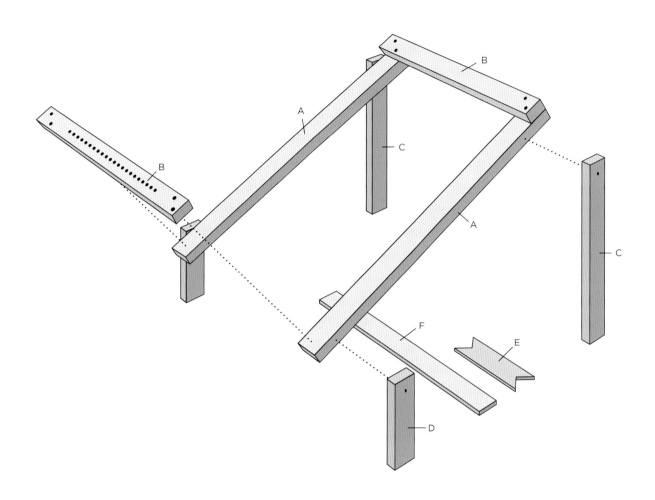

CUTTING LIST

KEY	NO.	PART	DIMENSION	MATERIAL
A	2	Frame sides	¾ × 1½ × 24"	1 × 2
B	2	Frame ends	¾ × 1½ × 18"	1 × 2
C	2	Back legs	¾ × 1½ × 12"	1 × 2
D	2	Front legs	¾ × 1½ × 5"	1 × 2
E	1	Shuttle	¼ × 1½ × 6"	Mull strip*
F	1	Shed stick	¼ × 1½ × 18"	Mull strip

*Can also be made from a paint stir stick.

TOOLS & MATERIALS

Measuring tape
Cordless drill and bits
Handsaw or utility knife
1½" finish nails
2½" machine screws and matching wing nuts (10–32)
Rubber non-slip furniture leg pads
Construction adhesive
Eye and ear protection
Work gloves

The Beginners Guide to Self Sufficency Projects for the Home

HOW TO BUILD A FRAME LOOM

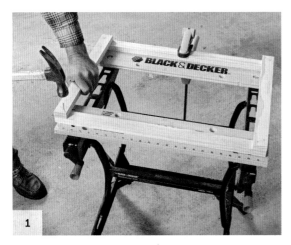

1. Set the frame sides flat on a clean, level work surface and align the frame ends across either end of the sides. Drill pilot holes and nail the ends to the sides using two finish nails at each corner. Check for square as you work and adjust as necessary.

2. Mark and drill holes through the tops of the four legs and through each frame side. The legs are attached 2" in from each end of the frame. The holes need to be the same diameter or just slightly bigger than the machine screws you've selected. After the holes are drilled and tested, mark the legs for each side and remove the bolts.

3. Mark the guide nail locations along the bottom end piece (the end with the short legs). The first nail should be positioned 3" in from the end, with nails every ¼" along the face of the end piece. The last nail will be located 3" in from the opposite end. Mark all the locations, then predrill the holes so the wood doesn't split. Drive a finish nail at each point, halfway into the wood.

(continued)

Homestead Amenities

HOW TO BUILD A FRAME LOOM (continued)

Drill a ⁵⁄₃₂" hole in the shuttle face, ¾" from the end, and centered between the edges. Repeat on the opposite side. Mark lines from the corners to the hole, and use a jigsaw or handsaw to cut a triangle from the end of the shuttle. Repeat on the opposite side.

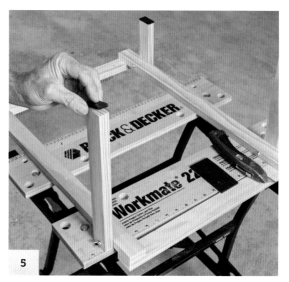

Attach the legs to the loom frame with the machine screws and wing nuts. Cut the rubber pads to the fit the bottom of the legs and glue them in place with construction adhesive. These will keep the frame from sliding when you are weaving.

WEAVING TERMS

There's a whole language around working with a loom. Once you learn them, the terms are pretty common sense, and they're necessary if you ever graduate to a more complex freestanding loom.

Warp: This is the main set of threads or yarn fibers that run vertically up the loom. The word is also used to describe the process of attaching these threads.

Weft: The yarn or fibers that run horizontally between the warp threads.

Heddle: To make things easier, avid weavers use this tool to maintain a space between alternating warp threads. This makes it much easier to weave the weft in and out of the warp. The project described here uses a shed stick for the same function.

Shuttle: You can weave the weft by hand, or do what most weavers do and wrap the yarn or fiber around this handy tool. It's basically just a flat stick with grooves or cutouts, around which the yarn is wrapped. The shuttle is then passed through the warp to quickly create the weave.

Shed: The space you create with a heddle or spacer stick—through which the shuttle and weft yarn passes—is called a shed.

USING YOUR LOOM

The basic idea behind this loom—and any loom for that matter—is to interweave yarn in perpendicular directions to create strong textile panels. Although yarn is the most commonly used fiber, other fibers can be used for different purposes, such as weaving placements for a dining table. In any case, begin by tying one end of the yarn around the bottom end (the end with the nails) of the frame. Secure it with a double slip knot and wind the warp under the top end, over the top end, then under the bottom end and back over, so that the warp looks like a figure 8. Use the brads as guides to create regular spacing between the strands of yarn. Slide the shed stick between the warp near the top end of the loom, then move it down toward the bottom end and turn it on edge, creating a space for the shuttle to pass through. Tie the end of the weft to the side of the frame at the bottom, then pass the shuttle through the warp and tamp the weft down to the bottom. For the next weft, slide the shed stick from the other side, over and under the opposite warps. In other words, where the weft went under a warp, it should now go over it. This can be painstaking work if the warp goes through every nail, but can go much quicker if the warp is spaced more widely. After the shed stick is all the way through, turn it on edge again and pass the shuttle back through. Then pull the weft tight, push it to the bottom, and repeat the first step. Work back and forth to create the weave, tamping down each weft row with the shed stick, so that it's snug to the row beneath it. (There are specialized tools called "beaters" for this purpose, but on a small loom like this one, the shed stick works just as well.) When you're done, untie or cut the warp at either end, and either leave the loose strands, trim them, or finish the piece with a fabric edge band.

20 | Manual Laundry Washer

Today's high-efficiency washers and driers are marvels of modern technology. They use less electricity and water than their predecessors and include more features than ever. However, for all their improvement on past models, they still use a significant amount of energy (especially when someone does a small, "quick" load—you know who you are), as well as a good deal of precious water.

What's more, washing clothes hasn't changed. The whole idea remains astoundingly simple: agitate the garments in soapy water for a period of time until dirt loosens and releases from the clothes. That basic function doesn't necessarily need a high-tech solution. If you're looking for something more sustainable, more environmentally friendly, and cheaper, you've found it—a simple, manual laundry machine.

This unit is not a challenge to build. All you need are a few basic carpentry tools, a few pieces of hardware, and a couple hours to assemble it. You can build it from components you'll find at any well-stocked hardware store or large home center.

The capacity probably won't rival the drum for your current washer, but is certainly large enough to handle most loads of laundry or animal blankets.

Once you've put your washer together, actually doing a load of laundry entails plunging the handle repeatedly for about 20 minutes. The machine makes great use of the lever principle so anyone—even someone with moderate strength and stamina—can easily wash a load of laundry by hand. Use biodegradable soap and you can merely empty the bucket in your landscaping as gray water. Hang the clothes up to dry (use the Clothesline Trellis on page 70) and you're done!

Save electricity, get a little exercise, and conserve water with this handy manual clothes washer. The ingenious lever action ensures that clothes are properly agitated and thoroughly washed.

BUILDING A MANUAL LAUNDRY WASHER

TOOLS & MATERIALS

Cordless drill and bits
Table saw or circular saw
Jigsaw
5-gallon plastic bucket with lid
(4) 2½ × ⅝" corner braces
(1) ¼ × 6" machine screw and nut or wing nut
(1) ¼ × 4" machine screw and nut or wing nut
Deck screws 2½", 3", 3½"
¾" stainless-steel wood screws for corner braces
¾" stainless-steel bolts, nuts and washers for pail lid agitator
Large bucket or tub
Clamp
Chisel
Sander
Utility knife
Eye and ear protection
Work gloves

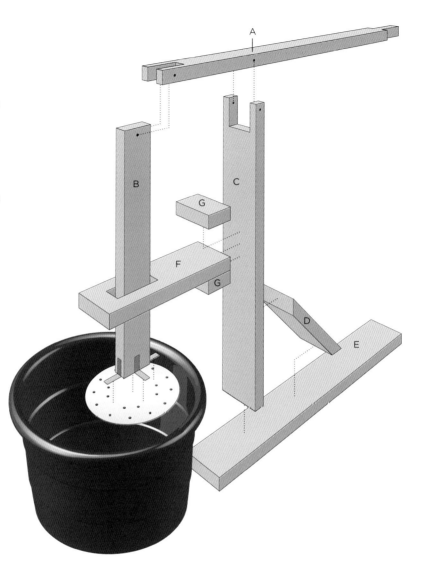

CUTTING LIST

KEY	NO.	PART	DIMENSION	MATERIAL
A	1	Handle	1½ × 3½ × 51"	2 × 4
B	1	Plunger	1½ × ½ × 25"	2 × 4
C	1	Support	1½ × 5½ × 36"	2 × 6
D	1	Brace	1½ × 5½ × 15"	2 × 6
E	1	Base	1½ × 5½ × 40"	2 × 6
F	1	Guide	1½ × 5½ × 18"	2 × 6
G	2	Cleats	1½ × 3½ × 5½"	2 × 4

HOW TO BUILD A MANUAL LAUNDRY WASHER

Cut the slot in the end of the handle to accept the plunger by making multiple passes with the table saw. Clamp a stop to the fence so that all the cuts are the same length. The slot should be cut centered on the face of the handle, 1½" wide by 4" deep. Clean the slot out with a chisel after cutting it and sand smooth. Cut and sand the opposite end of the handle to make a grip that's 1½" wide by 5" long. (If you don't have a table saw, just cut the slot with a jigsaw and square the end with a chisel.)

Cut the plunger to length. Mark a hole for the plunger at 1¾" from the side and 1" from the end so that it lines up with the end of the handle and extends ¼" beyond the top of the handle. Drill a ¼" hole centered in the side of the handle slot, 1¾" from the end. Continue the hole through the plunger and opposite slot. Use a drill press if you have one to get a straight hole; otherwise, mark and drill the holes in both sides of the handle, run the bit all the way through to straighten the hole, then drill through from alternate sides of the handle through the plunger. (If the hole is angled too much it may cause the plunger to bind when you move it up and down. If this happens, just redrill the holes with a ⁵⁄₁₆" bit.)

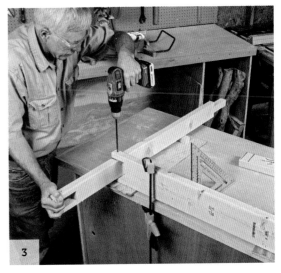

Mark and cut out the slot in the end of the handle support board. The slot should be centered on the face of the support, 3½" wide by 3" deep. Drill the ¼" pivot hole through the sides of the slot ¾" from the end and ¾" from the side. Drill the matching hole in the handle at 13¾" from the center of the hole plunger hole.

(continued)

HOW TO BUILD A MANUAL LAUNDRY WASHER (continued)

Attach the base to the opposite end of the support with 3½" deck screws. Miter the ends of the brace 45° and fasten it to the base and support with toenailed 2½" deck screws.

Mark the guide hole for the plunger arm in one end of the guide board. The outer edge of the hole should be 2½" from the end of the board. The hole will be a rectangle, 2 × 5½", with the long sides parallel to the long sides of the guide. Center it on the face of the guide board and drill holes at all four corners. Use a jigsaw to cut out the hole.

Screw the guide to the support, on the opposite side from the brace, using 3½" deck screws. It should be positioned 18" up from the base. Screw one cleat above and one cleat below the guide, snug to it, using 3½" deck screws. Also screw the support and braces together for more rigidity.

7 Attach the handle to the support with a ¼ × 6" machine screw and bolt or wing nut. Use washers on either side. The bolt should slide through the hole in the handle fairly easily. If it doesn't, run the drill bit through a few more times.

8 Attach 2½ × ⅝" stainless-steel corner braces to the sides and faces at the bottom end of the plunger. Screw the braces in place with ¾" stainless steel wood screws. Use a utility knife to cut the pail lid along the inside seam, then remove the lip. Center the end of the plunger on top of the lid and mark the location of the holes in the corner braces. Drill the holes for the corner braces, and then drill a varied pattern of about 20 additional ⅜" holes spaced evenly around the lid. This will allow water to pass through when you are agitating the laundry.

9 Attach the pail lid to the end of the plunger with ¾" stainless-steel bolts, washers and nuts. Slide the plunger arm up through the guide hole in the guide. Secure the tongue at the end of the plunger in the handle groove using a 4" machine screw, a bolt or wing nut, and washers. Check the operation of the handle and plunger and adjust as necessary. You can use a 5-gallon bucket for washing small loads or a larger tub for dropcloths, horse blankets, and other large items.

Homestead Amenities

Resources

AEE Solar
800-777-6609
www.aeesolar.com

Atkinson Electronics
800-261-3602
atkinsonelectronics.com

The Barefoot Beekeeper
www.biobees.com

Credits

Crystal Liepa: 66 (top)

Shutterstock: 6 (top right), 8, 13 (all), 24, 26 (both), 27 (both), 28, 62, 64, 65, 69 (both), 78, 83, 94, 96, 98, 100, 103, 104, 106 (right), 120 (right, both), 118

iStock: 122 (left)

Index

Acidity, home canning and, 104
Aerobic microbes, 38
Air, compost and, 38
Air-drying foods, solar dryer for, 109–111
Algae, rain barrels and, 52
Alpacas, 27
Animals, large farm, 25–27
Apples
　preservation methods for, 102, 103
　solar dryer used for, 109
Arroyo, 57–60
Asparagus
　plant compatibility, 82
　preservation methods for, 102

Bananas, solar dryer used for, 109
Bark chips, as mulch material, 83
Beans
　growing in containers, 68
　plant compatibility for, 82
　preservation methods for, 102
Bees and beekeeping, 29–33
　benefits of, 29
　five ways to keep safe and healthy, 30
　top-bar hive for, 30–33
Beets
　grown in containers, 68
　plant compatibility, 82
　preservation methods for, 102
Bok choy, 68
Botulism, 104
Boulders, in arroyos, 59, 60
Box cookers, 122
Broccoli
　growing in containers, 68
　plant compatibility, 82
　preservation methods for, 102
Broilers (fryers), 13
Brooder box, building a, 21–23
Brussels sprouts
　growing in containers, 68
　preservation methods for, 102
Bucket, manual laundry washer built with, 136
Building projects
　arroyos, 58–60
　brooder box, 22–23
　chicken coop, 14–19

clothesline project, 72–75
cold frame, 91–93
frame loom, 130–133
manual laundry washer, 136–139
pallet planter, 77–79
raised beds, 84–87
root vegetable rack, 114–117
solar dryer, 110–111
swales, 61
top-bar hive, 30–33
Bush beans, 82

Cabbage
　growing in containers, 68
　plant compatibility, 82
　preservation methods for, 102
Canning
　botulism and, 104
　Canned Food Safety Quiz, 105
　foods that are good for, 102
　process, 104–105
　pros and cons of, 101
　tools for, 104
　two main tools for, 104
Canning jars, 104
Cantaloupe, 82
Cardboard, in compost, 39
Carrots
　growing in containers, 68
　plant compatibility, 82
　preservation methods for, 102
Cauliflower
　plant compatibility, 82
　preservation methods for, 102
Cedar wood
　for chicken coop, 15
　for compost bin, 40
　for raised beds, 81, 86
　for solar cooker, 110
　for top-bar beehive, 31
Celery
　plant compatibility, 82
　preservation methods for, 102
Chafing dish, drying foods using a, 106
Chandler, Phil, 30
Chard, 68
Cherries, preservation methods for, 102
Chicken coop, building a, 10–11, 14–19

Chickens, 9
　breeds, 13
　broiler weight, 13
　building a coop for, 10–11, 14–19
　buying as newborn chicks, 12
　dual-purpose breeds, 13
　getting permissions from local municipality for, 12
　laying breeds, 13
　meat breeds, 13
Chicken wire, for chicken coop, 15, 19
Chicks
　brooder box for, 21–23
　care of, 12
　purchasing, 12, 13
Chinese cabbage, 68
Chromated copper arsenic (CCA), 81
Clothesline trellis, 71–75
Coal ash, avoiding in compost, 39
Coffee grounds, in compost, 35, 39
Cold frames, 63
Companion planting, 83
Compost(ing)
　bees and mixing your own, 30
　benefit of, 35
　materials for, 39
　mixing with soil, 83
　variables for, 38
Compost bins
　building a compost bin, 40–43
　building a two-bin composter, 45–49
　two basic kinds of, 37
Compound miter corner cuts, 125
Containers
　for plants, 68, 69
　for seedlings, 65, 66
　for storing dried foods, 107
Cooker, solar, 121–127
Corn
　growing in containers, 68
　plant compatibility, 82
　preservation methods for, 102
Cornish hen, 13
Creosote, 81
Cucumber
　growing in containers, 68
　plant compatibility, 82
　preservation methods for, 102

Dehydration. *See* Drying/dehydrating foods
Dehydration, preserving foods by, 101, 102, 103, 106–107
Drying/dehydrating foods
    checking to see if foods are dry when, 106–107
    dehydrator used for, 106
    drying times, 106
    foods that are best for, 102
    oven used for, 106
    pasteurization of foods and, 107
    pros and cons of, 101
    rehydrating foods after, 107
    solar dryer for, 109–111
    storage of foods after, 107
    treating foods before, 103
Drying, preserving foods by, 101, 102, 103
Dry streambeds (arroyo), 57–60

Eggplant, 68, 82
Eggs
    chicken breeds and, 13
    collecting from the nest, 12
Eggshells, in compost, 39

Farm animals, 25–27
Fencing, for large farm animals, 26, 27
Fertilizer, 26, 35
Fiberglass mesh/window screening
    for rain barrel, 52, 53
    for solar dryer, 110–111
Filters, rain barrel, 52
Floral bouquets, in compost, 39
Flowers, planting bee-friendly, 30
Fluorescent lighting fixtures, 65
Foil, for solar cooker, 121, 123, 126
Food preservation and preparation
    benefits of, 99
    by canning, 101, 102, 104–105
    choosing methods for, 101–102
    by drying/dehydrating, 101, 102, 103, 106–107
    by freezing, 101, 102, 103
    root vegetable rack for, 112–117
    solar cooker for, 121–127
    solar dryer for, 108–111
Food scraps, in compost, 39
Frame loom, 128–133
Freezing fruits and vegetables, 101, 102, 103
Fruit pits/seeds, avoiding in compost, 39

Game hens, 13
Garden(ing), 63–93. *See also* Compost(ing)
    benefits of, 35
    clothesline trellis for the, 71–75
    cold frame for, 89–93
    companion planting for, 83
    container types for, 69
    healthy soil for, 83
    pallet planter for, 77–79
    plants to grow in containers, 68
    raised beds for, 81–87
    soil for, 35
    starting seedlings for the, 65–67
Garlic, 82
Germination, 67
Goats, 9, 25, 26
Grass clippings, for compost, 38, 39
Green beans, 102
Gutter downspouts
    rain barrels and, 52, 55
    swales and runoff from, 61

Hay
    in compost, 39
    goats fed, 26
    as mulch material, 83
    as pig bedding, 26
Headspace, for canning, 104–105
Heddle, 132
Hens, collecting eggs from a brooding, 12. *See also* Chickens
Herbs, in plantar pallets, 78
Heritage chicken breeds, 13
Homestead amenities, 119–139
    frame loom, 128–133
    manual laundry washer, 134–139
    solar oven, 121–127
Honey, 9, 29

Kale, 68
Kitchen scraps, in compost, 37, 38, 39
Kits
    rain barrel, 52
    raised bed, 85
Kohlrabi, 68

Landscape fabric
    for arroyos, 59
    lining planters with, 69
    for raised beds, 84, 87

Laundry washer, 134–139
Lawn clippings
    for compost, 39
    as mulch material, 83
Laying breeds, chicken, 13
Leaves, shredded
    in compost, 39
    as mulch material, 83
Lettuces
    compatible plants, 82
    growing in containers, 68
    for pallet planter, 78
Lima beans, 102
Livestock, 25–27
Livestock manure, 37, 39
Live storage (cellaring)
    foods that are good for, 102
    pros and cons of, 101
Lumber. *See also* Pine wood; Plywood
    for chicken coop, 15
    for clothesline trellis, 72
    for compost bin, 40
    for frame loom, 130–133
    for manual laundry washer, 136
    for raised beds, 81, 84, 86
    for solar cooker, 110, 121, 123
    for top-bar beehive, 31

Manual laundry washer, 119, 134–139
Meat breeds, chicken, 13
Milk, from goats, 9, 26
Miter corner cuts, 125
Mosquitoes, rain barrels and, 52
Mulch, in garden soil, 83
Mustard greens, 68

Native plants, 60
Newspaper (shredded)
    in compost, 39
    as mulch material, 83

Onions
    growing in containers, 68
    plant incompatibility, 82
    preservation methods for, 102
Oven, drying foods in, 106

Panel cookers, 122
Paper towels/napkins, in compost, 39
Parabolic concentrators, 122

Pasteurization of foods, dehydration and, 107
Peaches, 102
Pears, 102
Peas
  growing in containers, 68
  plant compatibility, 82
  preservation methods for, 102
Pentachlorophenol, 81
Peppers
  growing in containers, 68
  preservation methods for, 102
  solar dryer used for, 109
Pet waste, avoiding in compost, 39
Pigs, 25, 26
Pine wood
  root vegetable rack made from, 113, 114
  for solar oven, 123
  for two-bin composter, 46
Plants. *See also* Food preservation and preparation; specific names of plants
  in arroyos, 60
  for containers, 68
  grown from seeds, 65–67
  for pallet planters, 78
  in rain gardens, 60
  vegetable plant compatibility chart, 82
Plexiglas
  for chicken coop, 11, 19
  for cold frame, 91–93
Plywood
  for brooder box, 22
  for chicken coop, 15
  for cold frame, 91
  for solar cooker, 121
  solar cookers built out of, 121, 123–127
Potatoes
  plant compatibility, 82
  preservation methods for, 102
Pots, garden, 69
Potting soil, 30, 39, 78, 83
Pressure canner, 104
Pulleys, for clothesline trellis, 75
Pumpkins
  plant compatibility, 82
  preservation methods for, 102

Radishes
  compatible plants, 82
  grown in containers, 68
  preservation methods for, 102
Rain barrels
  installing, 54–55
  purchasing and making, 52–53
Rain garden, 60
Rainwater, 35
  channeling, 57–61
  collecting in rain barrels, 51–53
Raised beds, 63
  benefits of, 81
  building, 84–87
  building from a kit, 85
  how to build, 84
  lumber for, 81
  with removable trellis, 86–87
  sizes and position of, 81
  watering, 83
Regulations, on keeping chickens, 12
Rehydrating foods, 107
River rock, in arroyos, 57, 58
Roasters (chicken), 13
Roost, for chicken coop, 11

Sawdust, in compost, 39
Seedlings
  four stages in growth of, 67
  growing mediums for, 66
  hardening off, 67
  requirements for growing, 65
  starting, 67
  transplanting, 67
Self-sufficient life, building a, 7
Self-watering containers, 68
Shed, 132
Sheep, 25, 27
Shelter
  farm animals, 25
  for goats, 26
  for pigs, 26
  for sheep, 27
Shuttle, weaving loom, 132
Soil, garden
  importance of healthy, 33
  for the raised bed, 81, 83
Soil test, 83
Solar cookers, 121–127
  building instructions, 123–127
  types of, 122
Spigot, rain barrel, 52, 53, 54, 55
Spinach
  growing in containers, 68
  preservation methods for, 102
Sprinklers, 83
Squash
  growing in containers, 68
  plant compatibility, 82
  preservation methods for, 102
  solar dryer used for, 109
Steppers, for arroyos, 58, 59, 60
Stones, for arroyos, 57–61
Storage
  of canned foods, 105
  of dried foods, 107
  root vegetable rack for, 112–117
Straw
  in compost, 39
  as mulch material, 83
Strawberries
  for pallet planter, 78
  plant compatibility, 82
  preservation methods for, 102
Subsidies, rain barrel use and, 51
Sunlight
  drying foods and, 103
  for growing seedlings, 65
  Solar cooker and, 121–127
Swales, 61

Temperature
  for a chest freezer, 103
  in cold frames, 89
  composting and, 38
  for food drying, 103
The Barefoot Beekeeper, 30
Tomatoes
  growing in containers, 68
  plant compatibility, 82
  preservation methods for, 102
  preserving, 99
  solar dryer used for, 109
Top-bar hive, 30–33
Trash can, rain barrel made from, 52–53
Trees, planting, 95–97
Trellis(es), 63
  building a raised bed with removable, 86–87
  clothesline trellis, 71–75
Turnbuckles, for clothesline trellis, 75
Turnips, 68, 82

Veterinarians, 25

Index 143

Warp threads, 129, 132
Washer, manual laundry, 135–139
Water(ing)
    compost and, 38
    raised beds and, 83
Water bath canner, 104
Weaving loom, 129–131, 133
Weaving terms, 132
Weeds, avoiding in compost, 39

Weft threads, 129, 132
Windbreaks, 95–97
Window sash, for solar dryer, 110–111
Wine barrels, planter box made from, 69
Wood ash, in compost, 39
Wood chips
    in compost, 39
    as mulch material, 83

Wool
    from alpacas, 27
    from sheep, 27

Yard debris, in compost, 39

Zucchini, 68